Our Climate Our Future

a comprehensive guide to the environmental crisis

JONNY FRIEND

Our Climate Our Future

First published in 2022

ISBN 978-1-8383722-9-3

Quirky Bird Books

Independently published with the support of
TAUK Publishing

Dedicated to Jack & Billy

CONTENTS

ACKNOWLEDGEMENTS

There are many people without whom the writing of this book just would not have been the same. The support they have provided has come in many forms, be it inspiration, motivation, encouragement, technical expertise, cover illustration, design, proofreading, listening to my rambling or giving me the time and space required.

Whilst each of those who have supported me have done so for altruistic reasons and expect no praise themselves, they very much deserve it;

Jack and Billy Friend, my two sons, you are the lights in my life. Without you I would never have had the strength to dedicate the time, energy and effort required in such an undertaking. I love you boys. You inspire me every day!

Sustainability expert Dave Knight provided encouragement and expertise, especially in the early days. Holly McKay read drafts of the first chapters early on and her comments and support were invaluable. Mio Hirayama also read chapters and encouraged me to carry on. Rose Tinsley proofread early chapters and gave me support. Emma Davies and I had various discussions on food and farming, leading to additions in the book. Anna Brindle and I had many environmental discussions, especially on consumption and the clothing industry. Bo Brindle did an amazing job in putting the cover together and making the book look far more professional than I could ever have done. Mia Rivers produced the beautiful illustration on the cover, encapsulating each chapter of the book. Steve Morehen provided much guidance in proofreading each chapter. Any mistake within the book now is entirely down to me.

And then to the most important women in my life. Firstly, my mother, Kay Friend. Mum, you have always been such a source of inspiration for me. You also read over early chapters and encouraged me throughout, as you have my entire life.

And last, but certainly not least, my amazing wife, Anna Friend. From the very start you have been there for me; challenging, encouraging, supporting and giving me the time and space to just get on with it.

To each of those above I am eternally grateful. Without them this wouldn't have been the same book, nor such a rewarding experience for me. Thank you!

INTRODUCTION

"Change is coming whether you like it or not"[1]
Greta Thunberg

This simple statement encapsulates so much in the world today. Change is not only coming but happening right now. Exactly what that change will be is determined so much by our actions as human beings.

Let's be clear, as is explored further within this book, Climate Change is very much real. We understand the causes and know the solutions. We also recognise that, not only are we not doing enough to combat this change, but we understand why. And it is not just Climate Change; our environment is being damaged in so many ways, leading to the Environmental Crisis we find ourselves within today.

It is, therefore, vital that each of us strive to comprehend what the issues really are and what we can do about them. As part of this I hope that this book may play some part in supporting your own understanding and action, and I wish to thank you greatly for picking it up.

I should point out that I am not a working climate scientist. I do have a scientific background, with a degree in Astrophysics and almost thirty years of teaching science in schools. Within this role, as well as leading on environmental action in school, I often found myself being asked about all kinds of environmental issues. Within this I recognised two things; Firstly, I wanted to increase my own understanding to not only adapt my own actions, but to support the learning of others. Secondly, there is so much information and misinformation around. There are also many amazing books on various aspects on the Environmental Crisis but, maybe due to my own ignorance, I hadn't found one which covered all the main issues.

When lockdown hit us in March 2020 there was time to think further on this. It had become clear that many of us would benefit from a summary book on the whole Environmental Crisis. This is what I have attempted to do here. The aim has been to pitch the book so that it gives people an understanding of the current situation without getting into too much heavy science. There is a balance to be made here, as some is required. A conscious decision was also made not to include graphs and to keep numbers to a minimum as these would put off some people. If you are interested in looking at these further, then many of the references include such details.

The book is split up into nine chapters. The first eight of these each focus on the principal issues relating to the Environmental Crisis. These include Climate Change, Denialism, Energy, Food, Consumption, Reliance upon Technology, Biodiversity and People. Chapter nine contains a whole host of actions which we can carry out. Some important points will, undoubtedly, have been left out, but the aim is to inspire people to think about the natural world and our role within it.

It can be difficult to discern what is based upon scientific evidence from what, quite frankly, is not. The spread of misinformation, whether deliberate or not, feeds into the rise of anti-science, populism, nationalism, racism, and general lack of empathy for one another. Who is such separation serving? It is important to think carefully about who owns the information we receive.

We can choose to blame others, separate further from one another, get angry, and carry out actions that make the situation worse, or we can recognise that each one of us has a role in this. No single individual, group, or nation should take the blame. It is through coming together we can make the positive changes required, not only for the lives of future generations, but also for us here and now.

In some ways it can be argued that Climate Change provides us with the opportunity to re-evaluate ourselves and our relationships with one another, nature, and what we feel is truly important in the world. We have been bombarded from birth how our lives will be so much better if we would just consume one thing or another. This is despite evidence that such materialism is having a negative effect on our own happiness.

What we see through this book, time and time again, is that we are a fundamental part of nature, of this planet. It is through re-connecting ourselves with nature and learning from it that we can create the positive change required.

Let's be honest though, some parts of this book do not make pleasant reading. The Environmental Crisis is difficult to take in. But there is also a message of hope. There are so many amazing examples of positivity around the world, and you can be part of that. All is not lost.

Remember, it is our climate, our future, and there is a better way…

CHAPTER 1

WHAT IS THIS CLIMATE CHANGE ANYWAY?

Whilst many of us have heard of this thing called Climate Change, there is so much information, and misinformation, regarding it, that a useful starting point would be to understand what Climate Change actually is, and why it is occurring. Most people know that it's related to the average temperature of the Earth so let's start with that.

Up to around two hundred years ago the average global temperature had been fairly constant for several thousand years[1]. Since then, however, it has increased by around one point two degrees Celsius[2], with around two-thirds of this warming occurring in the last fifty years. Added to this, the ten warmest years since modern records began have occurred since 2005[3]. From this, we can clearly see that the average global temperature is indeed increasing. This rise in temperature is known as Global Warming. But what is causing it? And is it something we should be concerned about?

The Greenhouse

Global Warming is said to be due to the greenhouse effect. Let us then try to first understand what this is. In order to do so, let's first consider the greenhouse. It is essentially a structure, mostly made from a transparent material such as glass, which is used to maintain a higher temperature in order for certain plants to better grow. But how does it do this?

When visible light from the Sun hits the glass of a greenhouse much of it passes through. On the inside this light energy then strikes objects within the greenhouse, causing much of it to be absorbed. These objects later release this energy back out, but as infrared radiation, rather than visible light.

The glass walls of the greenhouse, whilst transparent to visible light, reflect much of this infrared back inside, causing it to stay within the greenhouse. This causes an imbalance, as less energy leaves the greenhouse per second, compared to how much enters. The effect of this, is that the temperature inside of the greenhouse increases.

Greenhouse Gases

The Earth's surface, rather than being surrounded by glass, is surrounded by an atmosphere of different gases. Visible light from the Sun passes through this atmosphere and strikes the surface of the Earth, where it is absorbed. Then, like in a greenhouse, the energy is re-emitted, but again as infrared. What happens to it next is the more interesting part as far as global temperatures are concerned.

Within the atmosphere, some of the gases act just like the glass, in that they absorb much of this infrared, thus stopping it from leaving. If this too, like the greenhouse, causes an imbalance, then the temperature of the Earth increases. Since these gases work in a such a similar way to a greenhouse they are known as greenhouse gases (GHGs).

It should also be noted here that our atmosphere is nowhere near as large as we think. If we imagine shrinking down the Earth to the size of an apple, then the height of the atmosphere would be no more that the thickness of the apple's skin! Therefore, any change that we, ourselves, make to the composition of the atmosphere, can have a significant effect on its whole.

But how do Greenhouse Gases Work?

All objects, including individual molecules, have a natural frequency. For example, if you push a swing then it moves back and forth with each swing taking a certain amount of time. If you adjust the physical swing, for example by changing its length, then each swing takes a different amount of time. This means that whilst different swings have distinct natural frequencies, each of them has one.

Now, normally if you are pushing someone on a swing,

then you do so every time it comes back to you. In other words, you are pushing at the same frequency as the natural frequency of the swing. This causes the swing to gradually increase in energy, and the swing gets higher and higher. If, however, for some reason, you decide to follow the swing more in its motion, and push it far more frequently, including when it is coming back towards you, then the swing doesn't get the same gain in energy. Similarly, if you don't push the swing very often at all, then again it doesn't gain such a large energy increase.

This phenomenon, known as resonance, is often seen in nature. When the frequency of a 'push' matches the natural frequency of the object, then more and more energy is absorbed. Like with the swing, the timing doesn't have to be perfect, but the closer the frequency of the push matches the natural frequency of the object, the easier it gains energy.

Infrared rays are waves and so oscillate back and forth as they move along. This puts a force on any particle it collides with. Atoms and molecules each have their own natural frequency. Whenever the frequency of the infrared wave is similar to this natural frequency, then the infrared, and therefore the energy, is absorbed.

You may now have guessed that the greenhouse gases have natural frequencies very similar to infrared waves. This means that the greenhouse gases in our atmosphere are very good at absorbing the energy from infrared. This energy will later be released, but this will be done in all directions, causing a significant proportion to go back towards the Earth, rather than into space. When this happens so much that there is more energy coming onto the Earth than leaving it, the Earth's temperature increases.

What would the Earth be like without Greenhouse Gases?

The greenhouse effect itself is well established and understood across the scientific community. But could these so-called greenhouse gases really have a significant effect on global temperatures? Well, the evidence, and the scientific understanding, tell us that they do.

As part of this it is important to realise that greenhouse gases are not all bad and indeed our very survival depends upon them. If there were no greenhouse gases within our atmosphere, then the average temperature of the Earth's surface would be around minus eighteen degrees Celsius[4]. That is far too cold for us. It is also way below the melting point of ice, meaning that most water on Earth would be frozen. There would, therefore, be very little opportunity for life on Earth. The reason why the average global temperature is more like fifteen degrees Celsius, is due to the presence of the likes of carbon dioxide. In other words, thank you greenhouse gases – without you we would not be here.

So, what happens if the amount of Greenhouse Gas increases?

Ok, so maybe greenhouse gases aren't so bad after all, so what's the problem? Let's now then look at what happens if the concentration of these greenhouse gases within the atmosphere increases.

When there are a greater number of greenhouse gas particles in the atmosphere, more infrared energy will be absorbed. This means that even more energy is trapped, causing the temperature of the Earth to build up.

What we know is that the concentration of greenhouse gases is indeed increasing, as well as the average global temperature. The science on this is well understood. It is this increase in greenhouse gases which is causing this change.

Which are the different Greenhouse Gases?

The two most abundant gases in the atmosphere, namely nitrogen and oxygen, cause very little greenhouse effect, and as such are not considered greenhouse gases. The main gases that do lead to the greenhouse effect, the so-called greenhouse gases, are carbon dioxide, methane, nitrous oxide, and water vapour, as well as a number of other manmade gases.

It is also important to recognise that not all greenhouse gases are equal. Different greenhouse gases have different natural frequencies to one another and so are better or worse at absorbing infrared. Now we will look at each in turn but before doing so, we need to understand the term 'global warming potential'.

Global warming potential compares greenhouse gases to carbon dioxide, in how potent a greenhouse gas it is. In other words, it tells us how much better that gas is at trapping heat in the atmosphere, compared to carbon dioxide. The time frame it is compared over is one hundred years. It is sometimes done over different time periods, but one hundred years is the standard. We should also add that this comparison is for equal masses of the different gases.

Carbon dioxide is defined as having a global warming potential of one. Methane is far better at trapping heat and has a global warming potential of twenty-eight. This means that over a one-hundred-year period, if the same mass of methane and carbon dioxide were introduced into the atmosphere, then methane would trap twenty-eight times more heat than the carbon dioxide over the next hundred years[5].

It should also be noted that the more likely a greenhouse gas is to react with other particles, the less time it will stay in the atmosphere. Some greenhouse gases will remain in the atmosphere for far less time than a hundred years. To determine their global warming potential, the amount of heat they trap within their short lifetime is compared to how much heat carbon dioxide will trap within the full one hundred years. Other gases, such as carbon dioxide, will remain in the atmosphere for far longer than one hundred years but here, only the amount of heat trapped within the one hundred years is considered.

Now let's consider the main greenhouse gases in turn.

Carbon Dioxide

When most people think of greenhouse gases, they think of carbon dioxide. This is the one which gets the most media attention, and rightly so. Whilst it only makes up a tiny fraction of the atmosphere, it is vital to life on Earth. Plants, for example, use carbon dioxide in photosynthesis. This makes it the basis for almost all ecosystems on the planet. If there was no carbon dioxide on Earth, there would be pretty much no life.

Human beings, like all animals, produce carbon dioxide within their bodies, in a process known as respiration. We then breathe it out, or rather breathe out more than we breathe in. This is all fine since the percentage of carbon dioxide within the air we breathe in and out is very low. At high concentrations carbon dioxide becomes toxic, but these are at concentrations way above those in the atmosphere. Volcanoes also release carbon dioxide in the atmosphere. All of these are natural processes.

The reason why carbon dioxide is the greenhouse gas that gets the most attention is because, since the beginning of the industrial revolution over two hundred years ago, it is the one that has increased most within the atmosphere, and therefore, has had the greatest impact on global temperatures. But why is this happening?

Fossil fuels are fundamental to our lives today. Modern life, for most parts of the world, is as it is due to the use of fossil fuels. They are burned at huge rates around the world, supplying us with the energy which we use to travel, generate our electricity, heat or cool our homes, extract materials, and so much more. It is important to recognise this. If fossil fuels had never been discovered, then other sources of energy would have been developed further, but at a slower rate than that of fossil fuels. We wouldn't have many of the things that people value today. But is there another side to them?

Fossil fuels, such as coal, oil and natural gas, have formed from dead plants and animals over millions of years. The carbon that was stored within those plants and animals has become the carbon within these fossil fuels. As we burn them, this carbon combines with oxygen in the air and forms carbon dioxide. This carbon dioxide is then released and ends up in our atmosphere, leading to Global Warming.

To make matters worse, the natural absorbers of carbon dioxide, namely plants and algae, are reducing. The destruction of forests, known as deforestation, along with forest fires, not only releases carbon dioxide into the atmosphere, but also means that there are then fewer plants to absorb it. We see a similar picture in the oceans, where plankton, which underpin much of ocean life, have decreased in number. This again means that less carbon dioxide can be absorbed.

In combination, this means that the concentration of carbon dioxide in the atmosphere is indeed increasing. Measurements have increased from about two hundred and eighty ppm (parts per million) at the start of the industrial revolution to four hundred and fifteen ppm today[6]. This means that the concentration has increased by almost fifty percent in the last two hundred years. This is obviously a concern, especially when we consider that before the industrial revolution, the concentration of carbon dioxide had remained relatively stable for thousands of years.

It is this rise in concentration of carbon dioxide that is having the greatest effect on the observed increase in average global temperatures. And it doesn't look like stopping anytime soon. All the evidence tells us that it is not only increasing further, but at an ever-higher rate.

Methane

Methane is a powerful greenhouse gas. It only remains in the atmosphere for twelve years, but it traps so much heat in that time that its global warming potential is twenty-eight. If we look over just a twenty-year period, then methane traps eighty-four times more heat than carbon dioxide! This means that, particularly in the shorter term, methane plays a significant role in the increasing temperature of our world.

Like carbon dioxide, methane is found as a naturally occurring gas. Before the industrial revolution there were seven hundred and twenty-seven ppb (parts per billion) within the atmosphere. Currently the concentration of methane is one thousand eight hundred and eighty ppb[7]. That is an increase of around one hundred and fifty percent! It is also higher than any other time in at least two million years.

Methane is released into the atmosphere through natural sources such as wetlands and termites but also human activities such as landfills and agriculture. This is particularly prominent from animals such as cows, and in the cultivation of rice. Also, any biological material that decays without oxygen, such as in landfill, also releases methane.

Nitrous Oxide

Another naturally found greenhouse gas is nitrous oxide. It is so good at absorbing infrared that it has a global warming potential of two hundred and ninety-eight! It also remains within the atmosphere for over one hundred years and can damage the Earth's ozone layer.

Before the industrial revolution there were two hundred and seventy-three ppb (parts per billion) within the atmosphere. Currently the concentration of nitrous oxide is three hundred and thirty-two ppb[8]. That is an increase of twenty-one percent. Nitrous oxide is released into the atmosphere through natural sources such as lightning, soil and oceans, but also in human activities such as the use of fertilizers and the burning of fossil fuels and biomass.

Water Vapour

Water vapour is often overlooked as a greenhouse gas, yet it is the one that actually traps more heat than any other. The focus is on the other greenhouse gases, since those are the ones which are changing the most and therefore contributing to the biggest increase in temperature.

Found as both moisture and clouds, human activities don't directly affect the amount of water vapour in the air and its global warming potential has not been determined. We do know however that as the temperature of the Earth increases more evaporation takes place. Warmer air also holds more water. This means that levels of water vapour in the atmosphere are increasing. This causes even more warming, thus amplifying the effect of the other greenhouse gases.

Purely man-made Greenhouse Gases

There are also a group of manmade greenhouse gases such as hydrofluorocarbons, perfluorocarbons, sulfur hexafluoride and nitrogen trifluoride. These have global warming potentials of hundreds to tens of thousands and will stay in the atmosphere for thousands of years.

These gases used in such things as aerosol sprays, refrigerators, air conditioning, solvents, fire extinguishers, electronics, pharmaceuticals, as well as in extracting aluminium and magnesium. The concentration in the atmosphere of these gases remains low, thankfully, but is nevertheless increasing.

Enhanced Greenhouse Effect

Ok, so it should be clear now that the concentration of the different greenhouse gases within our atmosphere is indeed increasing due to human activities. Since it is human caused it is often referred to as anthropogenic. We also understand the science on how these greenhouse gases increase global temperatures. It is safe to say then, like pretty much all climate scientists, that it is this enhanced greenhouse effect which led to the increased, and increasing, average temperature of Earth.

Effects of the Increase in Average Global Temperature

So, the average global temperature has increased by one point two degrees Celsius since pre-industrial times and we know the reasons why. But this increase doesn't sound like much at all. When I first heard this, many years ago, I thought it would be a positive. Where I lived wasn't particularly warm for most of the year, so maybe it would make things nicer? What real difference could just over one degree Celsius really make anyway? Even with the scientific models telling us it will increase further, it's not that much, is it?

In order to answer this properly, we should consider, not just by how much the temperature is increasing, but what the consequences of this are. It's important then, to not only recognise the scientific predictions for the future, but indeed the stark reality of effects we are seeing today, in our very lifetime.

Climate and Weather

At this stage it is useful to define climate and weather. Climate is the temperature, rainfall, wind, ice, waters, land and living organisms within in given place, over a long time period. Weather, on the other hand, is the changing day by day, or even within a day, temperature, clouds, precipitation, humidity and wind in a region.

Weather is notoriously difficult to predict more than a few days in advance. This means that it is very tricky to know the exact weather on a particular day. It is much easier, however, to know what the average will be over a prolonged period. There are many examples of this within science where it is extremely difficult, if not impossible, to predict what will happen to individual particles, systems or even people, but it is much, much easier to predict what will happen over time, on average, to a large number of them.

Climate Change

The term Climate Change then is the long-term changes in climate, i.e. the long-term changes in temperature, rainfall, wind, ice, waters, land and living organisms over a long time period. It would be useful then to look at each of these in turn to see how they are also affected.

Temperature

We noted earlier that the average global temperature is around one point two degrees Celsius higher than it was two hundred years ago. On top of that, the ten warmest years since modern records began have occurred since 2005. There can't really be any doubt that the average global temperature is increasing. But what are the effects of these higher temperatures?

First of all, let's recognise that we have been using the term average global temperature. The reason for this is that the temperature is increasing by different amounts in different parts of the world. For example, the Arctic is warming at a far greater rate than the rest of the world[9]. Indeed, in June 2020, the Siberian town of Verkoyansk within the Arctic circle, experienced a heatwave with a maximum temperature of over thirty-eight degrees Celsius[10]! This is way, way above what would be expected. Let's look at some other recent examples…

Globally, July 2021 was the hottest month since records began almost one hundred and fifty years ago[11]. It is normal for July to be the hottest month of the year, but nothing like in 2021. When we look at the situation in different regions across the world, we see that July 2021 was the hottest on record in Asia, the second hottest in Europe, and in the top ten in each of North America, South America, Africa and Oceania[11].

Getting a real grasp of this increase in temperature can be tricky, and so, in order to help, Climate Warming Stripes were created[12]. These are visual, easy to understand, representation of how the average temperature of the Earth has changed since 1850. Their use should be encouraged as a powerful tool in supporting our understanding of the changing temperatures on Earth.

We should also note that having a higher global average temperature, doesn't mean that it is always hotter than in the past. The increases we are looking at are in the trend in average global temperature over time. Within this, sometimes we are hit with particularly high temperatures in the form of heat waves. These can have devastating effects. The European heat wave of 2003, for example, killed over twenty thousand people, with around fifteen thousand of these in France alone[13]. In the UK it caused around two thousand deaths. Many livestock also died, and food prices increased. It also caused water levels to significantly decrease in rivers and lakes, as well as huge forest fires across several countries, including Portugal and Spain.

The 2021, the heat wave of northwest USA and southwest Canada[14] has been described as the most extreme on record in North America, with record temperatures twenty degrees Celsius higher than the norm. This led to hundreds of deaths, and thousands of hospitalisations. In Washington, for example, it was the deadliest weather event in history, with over a hundred deaths[15].

Now, whilst there have always been heat waves, what the science tells us is that such heat waves are now far more likely than they were a couple of hundred years ago, and that some heat waves would have been virtually impossible without human-induced Climate Change[16].

We should also note that there is great variation across the world. Over land, for example, the increase in temperature is notably higher than the aforementioned one point two degrees Celsius. The oceans on the other hand, whilst still increasing in temperature, have so far done so at a measurably lower rate.

It is not necessarily the temperature itself that is always the issue. A consequence of increased energy within the atmosphere is the effect that it has upon the seasons, rainfall, wind, sea currents and other phenomena. As temperature further increases, there will be increased changes across the whole climate. Let's have a look at each of these in turn.

Precipitation

At higher temperatures, water evaporates at a faster rate. This means that there will be an increase in the amount of water vapour in the air, leading to more rain clouds and therefore greater precipitation, i.e. rain and snow. Therefore, as the Earth's temperature increases, there will be an overall increase in how much it rains.

As the global rainfall increases, individual regions are affected in different ways. Some areas expect to receive far more rain than they do currently, whilst others will receive less. In general, those regions which are currently relatively wet are likely to increase in both frequency and volume of rainfall. These regions are therefore also more likely to experience flooding.

Relatively dry regions are predicted to become drier, due to a decrease in both frequency and volume of rain. Therefore, they are more likely to experience drought. Within this, it should be noted that some regions, whilst receiving an overall reduction in rainfall, will obtain much of their rain in very heavy bursts. This means that whilst at times they may experience drought, it can soon change to flooding.

There have been many examples in recent years of the increase in frequency and intensity of floods. In 2021 alone, a large number of places experienced this around the world. These include; Storm Christoph in northern England and Wales had half a month's rain in two days, flooding at least four hundred homes, resulting in around a hundred million pounds worth of damage[17]; in July Germany and Belgium experienced devastating flooding which led to over two hundred deaths![18]; in China, several extreme floods led to over three hundred deaths and around fourteen billion dollars' worth of damage[19]; Nigeria now experiences floods every year, with disastrous consequences for those who live there[20]; at least one hundred and fifty people were killed in floods, which led to landslides in India and Nepal[21]; thirty-six communities were affected by flooding and landslides in Costa Rica[22]; and all this alongside destructive floods in Pakistan[23], New Zealand[24], Austalia[25], New York[26] and London[27], as well as over three hundred and seventy-five thousand people being affecting by flooding across west and central Africa[28].

With the examples mentioned above, consider not just those directly affected by the flooding, but also those who lost loved ones, family, and friends. Many lost homes and livelihoods. Far from numbers in a news report, these people are human beings, just like us, with hopes, fears and people who love them.

Now, whilst there have always been floods, there is little doubt that they are becoming more regular[29]. Climate Change is not the only culprit here, as ineffective water management systems have increased the effects of the flooding. Some may even point out that some of these regions have not received increased total rain. Whilst this is true over the year, when a week's, or even a month's worth of rain falls within just a number of hours, the consequences are clearly devastating.

Ok, so flooding is getting worse in many places throughout the world. But, what about the regions receiving less rain than before? Well, one dire example is the ongoing situation in Syria. The devastating consequences of the civil war there, which began in 2011, have been well documented. What is less frequently mentioned, however, is that for three years before the uprising, Syria experienced severe drought. This had a disastrous effect on previously fertile land, which led to the displacement of around one and a half million people from rural Syria to the cities[30]. Whilst other factors also, clearly, played a significant role, there is little doubt that this crippling drought played a major role in the political unrest and disastrous war in Syria.

In Kenya, right now, over two million people are suffering from food insecurity due to drought[31]. The normal rainy seasons just have not brought the rain. This was so eloquently described in a speech by Elizabeth Wathuti, at COP26, which in my view was the most powerful, heartfelt speech of the entire conference[32].

2021 also brought many more other droughts across the world; around half of the USA experienced the like[33]; Brazil had its worst drought in over ninety years[34]; similar led to millions facing hunger in Angola[35]; and a twenty percent general decrease in rainfall in the last twenty years, and the driest summer in over fifty years, has led to overwhelming drought in Iran[36].

Again, whilst droughts have always occurred, Climate Change is causing their frequency, and severity, to increase. Indeed, the current drought in Madagascar, in which thousands of people are on the brink of starvation, is thought to be the world's first famine to be caused solely by Climate Change[37].

Another consequence of the increase in temperature and dry conditions in parts of the world is the escalation in wildfires. One well known example is the Black Summer fires of Australia in 2019[38]. Here the reduction in rainfall over several decades, and the driest, hottest summer on record, led to the largest forest fire ever recorded there. This led to over twenty percent of the temperate and mixed forests being burned, nearly eighty percent of Australians being affected, and the death of around a billion animals. The widespread images of burnt koalas are heart breaking indeed. Not only that, but the fires also emitted many hundreds of millions of tonnes of carbon dioxide into the atmosphere. Even for those who only care about money, it is estimated that these fires cost Australians many billions of dollars, both in terms of direct damage, and losses in tourism.

2021 also saw many other ruinous wildfires, including across Italy[39], Turkey[40] and Greece[41], which was famously documented in video from those evacuating via ferry on Lake Evia, where scenes were likened to a disaster movie[42]. And whilst California has regularly experienced such events, the fire season now is starting earlier, and ending later than in the past[43]. It has even led to destruction of almost twenty percent of the world's majestic giant sequoia trees within the last two years[44]. And this is a tree once considered virtually fire-proof. This tells us so much about the intensity of these fires compared to how they once were.

In spite of this, in many countries there has a deliberate misinformation campaign, designed to divert attention away from Climate Change, by blaming these fires on arsonists. The evidence, however, is just all too clear. How Climate Change is driving these events is all too well understood; the increase in temperature and drought in such areas, has created the ideal conditions for such infernos. Forest mismanagement may well play a contributary role, but it is clearly not the main driver of these occurrences.

The evidence is all too clear then, that Climate Change is causing, and will continue to lead to greater extremes in weather on Earth. This is not something which is just predicted to happen in the future, it is happening right now. Obviously, floods and droughts have happened throughout human history, but what is increasing is the frequency and severity in which they occur. This will just get worse as temperatures increase further. It may well be that no single flood. or drought, can be blamed solely on Climate Change, but the increase on how often they occur most certainly can.

It should also be noted here that whilst climate

scientists have predicted rising temperatures with a high level of accuracy, their models actually underestimated such a large increase in extreme weather events[45]. The models, therefore, are being adapted to take account of this new evidence. This is what happens in science; when new evidence is obtained, our understanding adapts, leading to improved models moving forward.

Wind

Whenever there are differences in temperature, a divergence in air pressure occurs. This causes air to be pushed from areas of higher, to lower pressure. It is this movement of air which we call wind.

As noted above, the increase in average global temperature leads to greater variation in local temperatures. This will, therefore, affect wind patterns. Some regions will experience more wind whilst others will receive less. These shifting wind patterns may bring colder air to traditionally warmer regions, or vice versa.

One example is with the jet streams; strong currents of wind many miles above the Earth's surface which move around the planet from west to east. These winds meander as they travel along, rather like a river. In the northern hemisphere the Polar Jet Stream helps separate colder air to its north from warmer air to its south. This jet stream can form buckles which trap warm, or cold, air in these regions. This makes it notably hotter or colder than it would otherwise be, and for extended periods. Indeed, many of the extreme weather events mentioned above are attributed to, at least in part, this meandering jet stream[46].

The most spectacular, and devastating, wind events are cyclones. Note that hurricanes and typhoons, are essentially the same thing but in different parts of the world. Cyclones form over tropical waters as the warm, moist air is fed by energy from the ocean. At higher temperatures, there is more energy which feeds into this system, leading to more powerful cyclones. This all makes sense, so let's look at the evidence on what is actually happening. Well, whilst there has been a notable increase in the frequency of cyclones since the 1980s, if we look back over the last hundred years, it seems that there hasn't been much change at all[47]. Maybe this is nothing to worry about then.

Well, what is clear is that there has been an observable shift in their locations, with the number of those in the North Atlantic and central Pacific increasing, whilst the number in the southern Indian Ocean and the western North Pacific have decreased[48]. Not only that, but the evidence is also telling us that the proportion of cyclones that are the more powerful Category Four and Five storms has increased. Indeed, these have gone up at a rate of around twenty-five percent per one-degree Celsius increase in Global Warming[49].

There have, sadly, been many devastating examples of powerful cyclones in recent years; Hurricane Katrina, which hit southern USA in 2005, led to over one thousand eight hundred deaths, and over a hundred billion dollars in damaged property[50]; in 2021 Cyclone

Ana ripped across Fiji, putting over ten thousand people in emergency evacuation centres. It also caused massive damage to buildings, infrastructure and crops[51]; in 2017, Puerto Rico was ravaged by both Hurricane Irma and Maria within just two weeks of each other. These killed almost three thousand people, destroyed water supplies, crops and homes[52]; also in 2021, Cyclone Seroja hit Indonesia and East Timor, leading to flooding and landslides which caused submerged homes and at least seventy-six deaths, with others possibly swept out to sea[53].

So, as we can see, cyclones bring vast amounts of wind and rain to affected regions. This leads to flooding, damage to structures and loss of human, animal and plant life. They also lead to the erosion of sediment, as well as ruining local fresh water, by bringing sea water to freshwater streams and lakes. Whilst no individual cyclones can be directly attributed to Climate Change, the overall increase in intensity of these storms can.

Ice

Ice has a melting point of zero degrees Celsius. Therefore, when the surroundings is higher than this, ice will melt. As the average global temperature increases, more of the Earth's ice is exposed to higher temperatures and so changes from ice to liquid water. This is basic science, but what has actually been observed?

Let's first look at the poles. Arctic sea ice, for example, has been lost at a rate of thirteen percent per decade[54] since 1980! There are, of course, slight variations year by year, but the overall pattern is clear. When we look at Antarctica, however, the opposite seems to be true. The extent of Antarctic sea ice has actually been increasing since this time[55]. So, what's going on? Well, first of all, we saw previously that whilst the average global temperature is increasing, there are variations in this around the world. This would fit with what we are seeing with sea ice too, since whilst the total sea ice across the world is decreasing, in parts of Antarctica, it is increasing. But maybe there is more to it than just this.

NASA's GRACE satellite, or rather pair of satellites, is able to measure variations in the force due to gravity across the Earth[56]. It can measure this across the entire Earth every thirty days, so will pick up on any localized variations. This 'gravity data' from GRACE actually shows that Antarctica has been losing ice every year since at least 2002, and at an accelerating rate[57]. But this doesn't match with the observations on increasing sea ice in Antarctica. So, what is going on?

Well, it comes down to the fact that ice can flow. The glaciers in Antarctica are melting and losing mass. As this is happening, glaciers are moving at an increased rate, with some moving down to the sea. As part of this, ice shelfs, which are the part of the glacier near the oceans, are collapsing into warmer water. In other words, whilst the area covered by Antarctic sea ice is increasing, taking Antarctica as a whole, its ice mass is definitely decreasing.

So, what about in other glaciers around the world? Well, it has been shown that across two hundred and twenty thousand glaciers studied, almost every glacier around the world is losing mass[58]. Some of the worse affected are in Alaska, Iceland, the Alps, the Pamir mountains, and the Himalayas. With so many communities around the world relying upon summer run off from these glaciers, their potential loss could be devastating.

When we look at the data on the Greenland Ice Sheet, we see that this too, is losing mass, and at an accelerating rate. Indeed, it is losing ice to such an extent that it is considered the greatest contributor to rising sea levels[59].

With rising global temperatures, this loss in ice is expected to continue further. Whilst some individual glaciers may have a small increase in mass over a short time, such as due to an increase in localized snowfall, the overall trend is clear; global glacier mass is decreasing and has been doing so for at least forty years. As the glaciers melt, more water enters the oceans, increasing sea levels.

A further consequence of the melting ice is the change in albedo. Albedo is the proportion of the heat and light which is reflected, rather than absorbed, by a surface. White surfaces, like ice, have a high albedo. This means that they reflect a high proportion of heat and light back towards space. As ice melts, however, it exposes darker substances such as the land or water. This, instead, absorbs much of the heat rather than reflecting it, causing the region to warm further. This is part of the reason why some colder regions, like the

Arctic, are experiencing greater increases in temperature than other places around the world.

Waters

We saw above that melting ice leads to more water in our seas and oceans. To understand the consequences of this, we should consider both melting sea ice and land ice.

Sea ice floats and has a lower density than water. When it melts the total volume of water compared to the volume of ice within the water remains about the same, so has no effect on sea level. Of course, it reduces the concentration of salts in the water, but has no real effect on sea levels.

When land ice melts however, all of that water was previous not in the ocean. This means that the amount of water in our oceans in increasing. This, of course, leads to an increase in sea level, as has been observed.

The other major factor leading to a rise in sea level is thermal expansion. The oceans store about ninety percent of the extra heat energy from the enhanced greenhouse effect. This clearly increases the temperature of the oceans. Also, as water increases in temperature it expands. Whilst a quirk of water means that this is not true from zero to four degrees Celsius, the average surface temperature of the oceans is significantly higher than this. This means that, overall, the oceans do indeed expand with increasing temperature. As the volume of water then increases, sea level rises. This rise in sea level is happening right now and the increase in water volume from the effects described above far outweighs any extra evaporation of water at higher temperatures.

The observed increase in sea level has been about twenty-one centimetres since 1880, with about a third of this occurring in the last twenty-five years[60]. This may not sound like a lot, but some small low-lying islands have already been lost[61]. Also, the number of low-lying coastal regions around the world have already seen the sea encroach, causing a loss of land. This increase in sea level also further amplifies the flooding due to increased rainfall and hurricanes.

Marine ecosystems also rely upon the dissolved oxygen that the oceans contain. Whilst the oceans contain only a tiny fraction of the oxygen within the atmosphere, its level has decreased by about two percent over the last fifty years[62]. This has happened because when the temperature of the water increases, the dissolved oxygen molecules gain energy and move faster. This makes it more likely for the oxygen to escape the water and go into the atmosphere. Variations in ocean current also affect how well the oxygen content is distributed.

Whilst the observed reduction in dissolved oxygen may not sound like much, there is significant variation across the oceans, with some areas having far lower oxygen content than others. And whilst there have always been these 'oxygen minimum zones', they are expanding[63]. This, in turn, has a significant negative impact upon the survival of many marine organisms in these regions.

We find that around thirty-one percent of the carbon dioxide that is released by humans into our atmosphere is absorbed by the oceans[64]. This means that the amount in our atmosphere, whilst still too high, is far less than is would be otherwise. The effect on the oceans, however, is that it makes the water more acidic, which in turn affects entire ecosystems.

Any increase in water temperature also reduces how well carbon dioxide can be absorbed and stored with the oceans. This means that as more carbon dioxide is emitted into the atmosphere, the ocean will become less able to remove it. At a high enough temperature, the oceans may even emit more carbon dioxide than they absorb.

The other major factor to consider with the oceans is the surface currents. Ocean currents are driven by a combination of variations in temperature, and salt content. As the temperatures change then, the ocean currents will also be affected. The increased fresh water from ice also reduces the local concentration of salt within the water, further affecting the currents.

Sea currents are hugely important. Not only do they bring warm or cold water to different regions, thus affecting the temperature and weather of those places, but they also lead to the movement of nutrient-rich waters from place to place. As these currents change, many areas will be affected, with potentially highly significant consequences.

The Seasons

Seasons are also affected as the global average temperature increases. For example, the number of cold winter days can decrease, and spring can appear earlier, and autumn later.

Whilst this may sound ok, many species rely upon the timings of the seasons for the different stages of their lives. As the seasons shift, so the species, and those that rely upon them are also affected. This includes laying of eggs, migrations, blooming flowers, and pollinators. Different species can also be affected in different ways, leading to some species becoming active when other species that they rely upon are not. More on this in Chapter Seven.

Living Organisms

The vast majority of species on Earth are affected and will continue to be so to a greater extent, by Climate Change. Each of the changes in the natural world mentioned above can affect their lives. Let's have a look at how some of these may affect life.

In 2021, thousands of baby flamingos were found dead at Lake Tuz in Turkey[65]. Waters had receded so much, that many of them just weren't able to travel from their nesting ground to the water, and so died. Not only had there been significant drought in the area, but dams had diverted water away from the lake for use in agriculture. The combination of these, both brought on by human actions, caused so many flamingo hatchlings to pay the ultimate price. Also in 2021, extreme heat in Canada is thought to have been responsible for the deaths of over one billion (!) marine animals within a week[66]! For example, the heat essentially cooked many mussels alive. There were also wildfires at the same time, causing untold damage to wildlife there too.

Chapter Seven discusses the effect on species in more depth, but let's be clear, the current extinction rate of species is estimated to be between one hundred and one thousand times the natural rate[67]. Some scientists even refer to our time as the Earth's sixth mass extinction of species. Human enhanced Climate Change is not the only cause of this, but it is a major factor. Other human causes of such extinction include habitat destruction, waste, and other pollutions.

Changes in the climate have, of course, occurred in the past, but the big difference is that they have never occurred to the extremes which we are currently seeing. This means that many species will be too slow to adapt to these changes in temperature and be forever lost. The species most susceptible to these changes are those which are highly specialised. Other species which are more adaptable and inhabit a range of environments are more likely to survive.

There are many ways in which the lives of different species may be threatened. An increase in temperature, for example, can directly make it too hot for species to survive. The increase in extreme weather events such as hurricanes, floods, droughts, and forest fires, also destroys the habitats upon which many living things depend. Also, as some species within an ecosystem die out, all others are affected too. Some may lose their food supply whilst others may increase greatly in number as their predators die. This can lead to plant pests increasing greatly in number, devastating the plant life around them, upon which the rest of the ecosystem is built.

We also know that plants use carbon dioxide for photosynthesis. As the concentration of carbon dioxide within the atmosphere increases, there is more available for plants to use. Also, at higher temperatures, as long it is not too hot, trees and other plants will be able to grow where they previously couldn't. Other regions, however, will become too dry or too wet. Overall, this will lead to ecosystems shifting away from the equator and towards the poles, and to higher elevations. Local ecosystems could be destroyed in some regions, as they are created in others. Overall, in time, there may be an increase in plants, but not everything will be able to adapt to the rate of shift and species that have survived for millions of years will become extinct.

Polar species, particularly those in the Arctic, which is all ice and sea, depend upon the ice for their survival. As it melts, and the temperature and salt concentration of the water change, food sources move on or die out. This means that the species that rely upon them will likely become extinct.

Coral reefs are well known for their beauty and high levels of biodiversity. Tiny coral polyps release calcium carbonate which, over thousands of years, give rise to structures, creating a coral reef. The many colours of these structures come from a variety of algae which live on them. Changes in temperature, pollution or light can make the polyps expel the algae, leaving the coral structures white. The coral is not necessarily dead at this point and can recover but the whiteness, or bleaching, is a sign that the coral is under stress and if the conditions do not improve the coral will die. If this happens the rest of the ecosystem will be adversely affected.

Each of the effects described above are already happening, and are predicted to increase in the future, as the average global temperature increases further. Whilst the measured increase of just one point two degrees Celsius over the last two hundred years doesn't sound like much, it is clearly having a significant effect on our climate and will continue to a greater extent in the future.

Feedback Systems

Earlier we explained that it is the increase in concentration of greenhouse gases, directly due to human activities, that have caused, and will continue to cause, the increase in average global temperature. We have also seen some of the other effects on our climate from this increase in temperature. But this is not the whole picture. There is also a combination of positive and negative feedbacks systems which feed into this all, which further amplify, or reduce, any effects.

Positive feedback systems heighten the effect of the increased concentration in greenhouse gases by further increasing the rising global temperature.

As mentioned previously, as the average global temperature increases so does the rate of evaporation of water. This increases the concentration of water vapour, a notable greenhouse gas, into the atmosphere. The consequence of this is to further increase the temperature, which in turn leads to ever greater evaporation, in a self-reinforcing cycle.

We have also noted that as ice and snow melts, darker surfaces below are exposed. This means that the albedo is reduced, and less heat and light is reflected back to space as more is absorbed, again increasing the temperature of the Earth.

Permafrost is land which is frozen below the surface for at least two years. Much of this has been frozen for far, far longer. The soil here can contain vast amounts of carbon left over from dead organisms in the past. As the increase in temperature causes this permafrost to melt, methane and carbon dioxide are released from the soil, into the atmosphere. Again, further increasing the average global temperature.

The increased occurrence of drought, alongside higher temperatures, increases the likelihood of forest fires. As the plants burn, more carbon dioxide is released into the atmosphere, and there are fewer such plants to absorb it. This, once more, further increases the concentration of carbon dioxide in the atmosphere and thus leads to even higher temperatures.

Thankfully, there are also negative feedback systems too, which reduce the effect of the increased concentration in greenhouse gases.

A combination of having access to more carbon dioxide, rainfall, and higher temperatures, means that new regions are opened up for trees and other plants to grow. These extra plants remove carbon dioxide from the atmosphere.

Warmer, darker bodies also emit more energy than cooler, light ones. As the Earth warms and darkens it radiates more energy from the surface. This reduces the ratio between how much energy is coming onto the Earth per second to how much is leaving. At a high enough temperature, this can lead to more energy leaving, rather than entering the Earth, thus reducing the Earth's temperature. This won't balance out until a notably higher temperature, however. The amount of damage done before we get to that stage is too great, so we want to avoid getting anywhere near there.

We also mentioned earlier that the amount of water vapour in the air increases with temperature. As well as the heating effect, this water vapour can also cause more low-lying clouds. These clouds reflect much of the heat from the Sun, having an overall cooling effect. In other words, increased water vapour can have both a warming, and a cooling effect on the Earth. It all depends upon where in the atmosphere it goes.

There are also some particles, namely aerosols, which act to reduce the average global temperature of the Earth. Aerosols are small particles such as dust, ash, soot, sulphates, and nitrates. Some of these are from natural sources, whilst others come from human activities. These aerosols reflect heat and light from the Sun back into space, thus reducing the amount of energy coming onto Earth, leading to cooling. Aerosols can also encourage water vapour in the air to form clouds, again causing a cooling effect.

It has been proposed that aerosols should be deliberately injected into the atmosphere to reduce warming, but the negative impact on living organisms means that this would be counterproductive.

Aerosols reduce air quality and cause damage to the lungs, as well as asthma and even lung cancer. Even if the aerosols are released high in the atmosphere, some would still make their way to the surface. I would suggest that the solution to Climate Change cannot be one that will cause so much damage to life.

What about Climate Change in the Future?

Any prediction of the future, by definition, has uncertainty within it. After all, the future has not yet happened. This does not, however, mean that we should ignore such predictions from scientific models, but we should recognise that this inbuilt uncertainty leads to a range of possible outcomes. Part of this is that we don't know the extent of future human activities. How much will we strive to mitigate Climate Change in the future? Will we develop the political will to carry out the required change? And if so, when?

To obtain as reliable a prediction as possible, scientists run models in super computers thousands of times. This allows them to adjust the concentrations of the different greenhouse gases and effects of feedbacks. The Fifth Assessment Report of the Intergovernmental Panel on Climate Change (IPCC), focused on four future scenarios[68]. These ranged from a relatively low to high emission scenarios.

It should be noted at this point that in whichever scenario projected, due the amount of greenhouse gases already without our atmosphere, all the models show that the average global temperature will continue to increase. The lowest emission scenario gives a total increase of one point eight degrees Celsius whereas the higher gives this increase as over four degrees Celsius.

The differences between these two scenarios may not sound significant, but each of the consequences of a rising global temperature mentioned earlier in this chapter depend greatly upon just how much the temperature goes up by. Every single increment in the increase of average global temperature will have serious implications for not only many species on Earth but the stability of our own lives. Extreme heat events, floods, droughts, sea levels rising, reduced food security, as well as habitat and biodiversity loss, will all become more severe with every increase in temperature. The increase of over four degrees Celsius over such a small time period would be devastating for much life on Earth as well as flooding vast coastal areas, greatly reducing food security leading to ever more famine, and significantly increasing the frequency of heat waves, hurricanes and drought. Let us not get to that!

In 2018 the IPCC released a special report on the importance of limiting warming to one point five degrees Celsius. It stated that at the current rate of increase, the average global temperature would reach such a temperature at some point between 2030 and 2052.

As part of this report, it stated that we have until 2030 to reduce carbon dioxide emissions by forty-five percent from 2010 levels if we are to limit the average global temperature to no more than one point five degrees Celsius above pre-industrial levels. The science tells us that the climate risks of going beyond one point five degrees Celsius, even by a relatively small amount, will be noticeably greater than if this number is kept to. This is what was meant when we were told that there were twelve years to make a difference.

The same report also tells us that carbon dioxide emissions need to reach net zero (i.e. as much carbon dioxide is drawn down from the atmosphere as is emitted) by 2050. But the Special report for one-point-five degrees Celsius, clearly states that Global Warming is likely to reach one-point-five degrees Celsius between 2030 and 2052 if it continues to increase at the current rate[69].

This means that significant changes need to be made now, and the greater the reduction in emissions, the lower the chance that we will not exceed this temperature. This is also why environmentalists, climate scientists and activists, find it difficult to hear when governments want to reduce emissions at a far lower rate. One is based upon what the science tells us, whilst the other is a political decision. We cannot continue to ignore what the science tells us.

But even this is not the final goal. As noted, there is already too much carbon dioxide in the atmosphere. Even if we became carbon neutral today, this so-called Legacy Load would still cause warming for decades, or maybe centuries to come, though obviously far lower rate. The goal then must be to become carbon negative, until such a time that the Earth is in balance once more.

The Sixth Assessment Report of the IPCC was published in August 2021[70]. To be honest, it's not pleasant reading. Its findings are clear; that Climate Change is already affecting every inhabited region across the globe with human influence contributing to many observed changes in weather and climate extremes.

It should also be noted that many climate scientists believe that there is little chance that we are going to be able to limit warming to one point five degrees Celsius. This may be seen as a reason to give up. If we are going to get so hot anyway then what's the point? But actually, the opposite is true. The further we go beyond this temperature, the worse it will be. Our aim must be to work together, as one people, as one planet, to mitigate Climate Change.

Summary

Climate Change is real, caused by human activities, and happening right now. It will also continue, and get worse into the future, unless we do something about it. This is what the evidence tells us, this is what the science tells us, this is what climate scientists tell us. It is already having, and will continue to have to a greater extent, a significant impact on life on Earth. But all is not lost, we can do something about it. Change is occurring, but we can affect by how much. In order to do this, we need to embrace the science, understand what changes need to happen individually, locally and globally, and then actually carry them out. Many people around the world are already doing this, but far more must come. We will look at what we can do both as part of a group, and as individuals, in Chapter Nine. The future is in our hands and the time to act is now. Remember, this is our climate, our world, and there is a better way.

CHAPTER 2

THE POWER OF DENIAL

Whilst the evidence on Climate Change is overwhelming, and the science is well understood, there are still many people in the world who deny its existence, or at least that we are its cause. The reasons for this are rich and varied, but it is important to look into the why, as well the evidence used. By doing this we can work together for a better future. If, instead, we refuse to address it, then this would be as bad as rejecting any other evidence that doesn't align with our world view.

Why Climate Change is Denied

Let's say it… Climate Change is scary. The reality of what is happening, and what will happen to a greater extent moving forward, is painful. For some the scariest part is that it tells us that the way most of us live our lives is unsustainable. It is not only others that need to change, but we need to make changes ourselves. It is therefore, far easier and far nicer, in the short term at least, to not believe in Climate Change.

But it is more than this. It is a natural reaction to protect ourselves, and different people do this in different ways. One way is to subconsciously close our eyes to the truth since it is too difficult to deal with. When this occurs, it is not possible to recognise that it is happening, as our subconscious acts to put up barriers to seeing the truth behind the evidence. Climate Change can also just feel wrong. It doesn't fit with our world view. It just can't be true. It is also hard to admit that we, or our loved ones, people whose opinions we respect, are mistaken.

Psychologically there are many reasons why we may deny Climate Change. Therefore, we cannot and must not blame others for their natural and understandable human reactions. Attaching blame creates a 'them and us' situation. Attributing blame and attacking people for such things can also cause people to go on the defensive and get so set in their views that there is no way out.

No, Climate Change is too big to cause such separation between people. What we need is to support each other and work together with everyone, whatever their views. Let us embrace all of humanity and help others understand that, whilst it is hard, together we really can make a difference. As part of this, we should explore different ways for people to see the truth, rather than live in denial.

Deliberately Spreading Doubt

There are, however, others that do understand Climate Change but deliberately misinform others about it. This is generally done in order to maintain short term gains, such as profits. This is especially true with large corporations, particularly those in the fossil fuel industry and those which fund them. Too often these are people who understand the science, but instead do everything they can to deny, or at least instil significant doubt about, human caused Climate Change.

Here we are talking about extremely powerful people and corporations. These are the ones who make the greatest profit from the current political and economic system. They also control most of the mainstream media and have a huge influence over the policies implemented by governments. These people are doing a very clever job of instilling doubt in people's minds. It has been said that some such people ignore Climate Change, but nothing could be further from the truth. Vast amounts of effort and expenses are used to deny it. If enough people recognised the severity, consequences and required actions of Climate Change then it would mean that their profits would diminish overnight.

This issue for such deniers is that the evidence that Climate Change is occurring and that this is affecting our planet is overwhelming. Other tactics are therefore required. Eminent climate scientist Michael E. Mann eloquently describes, as the four Ds, the varying tactics employed by such parties in their denialism[1]. The first is 'Delay'; where Climate Change is accepted to be real but that it's not really an issue; it is deemed that we can easily adapt and cope with it. The second is 'Division'; where tactics are employed to get climate activists to argue amongst themselves and to judge each other on their lifestyles. This is done to stop them acting as one, far more powerful, voice. The third is 'Deflection'; which puts the blame on individual behaviours rather than policy changes which would make a far greater difference. Note, as we will discuss in Chapter Eight, the most important individual actions are to become, or to put pressure on policy makers, whether through voting, or protest, or a variety of other actions. The fourth is 'Doomism'; which promotes the myth that it's too late, that there is no point in acting, since too much damage has already been done. Each of these are carefully designed to affect the mindset of the likes of you and I, to stop us bringing about positive change. We cannot allow that to happen.

So, is it now right to apportion blame to such deniers? I still say no, but we very much need to challenge what they are doing and continue to act in such a way as to bring about understanding to all. OK, so there may be some individuals that we will never get through to, but we can never know exactly who they are. And so, whilst I very much disagree with their actions, and will continue to fight against what they are doing, we must remember that we are all people. Often those who carry out such acts have been celebrated for their financial success, feel validated by it, and are living in a society where it has been reinforced from birth that material wealth is what we strive for. No, do not blame these people, but support them in grasping the big picture, help them see, however difficult it may be, that they too can change and play a positive role in the world moving forward.

Note that this does not, of course, mean that it is right to agree with those in denial. In the scientific world, if the evidence clearly demonstrates one thing to be correct, then opposing theories are simply wrong, and are therefore rejected. As more evidence is obtained, our understanding grows. Our thinking must remain based upon the evidence. But we must not cherry pick, distort, or misinterpret this evidence so that it shows us what we want it to. Such malign belongs to the realm of those who wish to not face the truth, whatever their reasons may be. Indeed, many reasons given by climate deniers have a small element of science at their core, but it is never the full story. Let us look at some of these examples now.

Reasons People give for Climate Denial

Very few people now completely deny that Climate Change is taking place. The evidence that it is happening is just too reliable, abundant, and explicit. There are some people however who still deny that Climate Change is caused by human activities. There are many theories, often based upon a small aspect of science, that it is not human activities, but natural causes, that are causing our climate to change. It is important then, to look at these in turn, so we can make informed decisions. Let the evidence be our guide.

Uncertainty

The Earth's climate is complex, with many interconnected components, affecting each other in many different ways. This is sometimes given as a reason to dismiss human caused Climate Change, stating that is too complex for us to understand, but this goes away from the very essence of science.

The aim of science is to understand the natural world. Scientists understand that uncertainty is a natural and fundamental part of the physical world. It's not just that we don't understand it well enough, uncertainty is part of the science itself. Science does not deal in absolutes.

Unfortunately, this is not always recognised outside the scientific community. Part of the problem is how science is taught in schools and how it is conveyed in the media. We are often told that the science is unquestionable and indisputable. This means that when the science goes against our world view it is easy for the uncertainties to be highlighted, because of course they are there. This is a sadly, often deliberate, way to make the rest of us doubt the science and instead follow a non-scientific viewpoint which is more in keeping with our ideas on how we want the world to be.

The media is also used cleverly to twist uncertainty to make it seem as if scientists aren't clear on Climate Change. If scientists aren't one hundred percent sure then surely, until certainty is obtained, it's ok to carry on as normal. We, therefore, don't need to make changes in our lives. Certainty, however, can never be achieved in any scientific field. This makes this argument extremely powerful in allowing people to maintain the status quo. But isn't this ok? As there is doubt then we can't know that human caused Climate Change is real, right?

Well let's look at the level of understanding. The amount of evidence gathered on Climate Change is staggering and quite possibly more than in any other scientific field in history. There is so much evidence that no single computer on Earth can deal with it all. Different computers need to work together, modelling different aspects and using different evidence. Climate scientists around the world develop and share computer models to best make sense of what is happening. These are brought together to make predictions for the future.

To monitor their accuracy, these models are also run from points in the past to the present day. Comparing these to the situation now suggests that these models are very accurate. When these models predict the future, they also show strong similarities. Variations do occur, in part at least due to differences in predicted human caused emissions, but when they are brought together the message is clear; human activity is causing Global Warming, which is changing the climate, and this will continue to happen. So yes, there is uncertainty in the climate models. And yes, they do predict differences in how much the average global temperature will increase in the future. But all the models show an increase in temperature, and it is this increase that we need to mitigate.

So, uncertainty is inherent within science and within climate models. It may be that the future global temperatures will be at the low end at the predicted values and so not as bad as is feared. Let us hope that's true. The change in the climate will still be significant, but far less than it may otherwise be. It may be however that it could be much worse. It could be that future temperatures will be at the top end of the predictions, meaning that the situation will be much worse than feared. Of course, since it is in the future, we don't know where within the range it will be. But let's say it again; it may be at the worst end of the uncertainty. This inherent uncertainty, therefore, doesn't tell us that we probably don't need to do too much, it tells us that we very much need to, and at a higher level than we might otherwise have thought!

Climate Scientists are Fixing the Data and Telling us Lies

It has been claimed that climate scientists are hiding evidence that the average global temperature is actually decreasing, rather than increasing. If this was true, then there must be a vast conspiracy between climate scientists across the world. This came to a head in 2009 when emails were hacked from the Climate Research Unit at the University and East Anglia[2].

The emails included a term "hide the decline" which was used to suggest that climate scientists were hiding a decline in the average global temperature. What was actually being discussed was a decline in tree growth at a number of high-altitude locations. Tree growth does generally increase with temperature. Whilst this data diverges with what would be expected with increasing temperature, there are other factors such as rainfall which could have reduced such growth. This divergence is well known, public and has been discussed in a number of scientific publications. No evidence was hidden. What we know is that the direct measurements of temperatures acted to hide the reduction in tree growth. It is this that was being referred to. In any case, any localised anomaly is no reason to deny the overwhelming evidence which supports the increase in average temperature on a global scale.

Interestingly, it has also been stated that climate scientists lie and conspire to get more funding, when so much of the information put out to suggest doubt about Climate Change comes from those who work for or are funded by the most profitable industry on the planet, namely the fossil fuel industry.

All of this is also not helped by media showing false neutrality. This happens when they put out opposite viewpoints on Climate Change, with climate scientists representing the established science and non-scientist climate deniers stating the opposite. Often this has been done as if each were as credible as the other, even though one is basing their statements upon huge amounts of research by thousands of scientists across the world and the other, quite frankly, is not.

This of course fits with the mainstream media who have denied or played down Climate Change for years. More on this in Chapter Eight, but mainstream media is owned by those who want to maintain the status quo;

a status which is not only challenged by Climate Change itself, but by how much society, the systems, the economy, and how we live our lives need to change if Climate Change is real and we are serious about mitigating it.

All this is designed, however, to put doubt in people's minds about whether human caused Climate Change is real or not. When doubt enters our minds then it is natural for many people to carry on their life as normal and not worry about it. In the short term at least, it a nicer, and easier way to live.

Over Thirty Thousand Scientists said it's not True

In 1998 the Global Warming Petition Project was launched, with the aim of collecting as many signatures of 'scientists' as possible. The petition stated that there is no convincing evidence that the human release of greenhouse gases will cause catastrophic heating of the Earth's atmosphere, and even that increased levels of carbon dioxide would bring benefits, rather than harm, to plant and animal life[3].

As of November 2021, just over thirty-one thousand had signed it. This would suggest that there must be issues with the evidence. So, what's going on? Well, when you look at who signed it, only around zero-point-one percent are purported to be actual climate scientists[4]. And it is not even stated who on the list these are. Indeed, anyone with a science degree can sign it. There are, for example, over three thousand medical doctors on the list. No one would go to a climate scientist with a medical condition since that is not their area of expertise. Similarly, how can a doctor claim to understand Climate Change better than an expert in that field?

Also, when the petition first emerged, it, and the accompanying documents, were made to look like official papers from the USA's National Academy of Science (NAS), even though they weren't. This was clearly a deliberate attempt to deceive others into thinking it was coming from a credible source of science. So much so, that the NAS themselves came out against it, and stated that the petition was a deliberate attempt to mislead others and was not based upon scientific evidence.

A simple search on the internet of many of the signatories comes up blank too. There is no scientific rigor here, and the petition is quite clearly not what it claims to be.

It's due to the Urban Heat Island effect

The Urban Heat Island effect is where a built-up area, such as a city, is noticeably warmer than the surrounding countryside. This comes from the extra energy released from human activities and from buildings, as well as the different surfaces that have been created[5].

It has been suggested that since such urban areas have expanded since the industrial revolution, the apparent increase in average global temperature is due to this

Urban Heat Island effect, and that apart from this the average global temperature hasn't increased.

Well, this is a bit disrespectful to scientists, who well know about, and understand, the Urban Heat Island effect. They obviously take this into account. Also, when data is obtained from remote rural locations these too show clear increases in temperature. The Arctic region, for example, which has no large urban areas, has experienced the largest increase in temperature. Therefore, the Urban Heat Island effect is not falsely suggesting a Global Warming, the warming is very much real.

Since we are on the topic of the Urban Heating Effect however, it seems prudent to make clear that it is in many large urban areas in which the biggest increases in temperature are occurring, and will continue to occur. This puts at risk millions of vulnerable people in cities across the world.

The Earth Cooled Mid-century

In the middle of the twentieth century the average temperature of the Earth showed an overall decrease by zero point two degrees Celsius from the 1940s to the 1970s. Whilst this is within the context of an overall increase in average global temperature during the twentieth century, it is important to understand what happened.

After the Second World War, industry increased to recover damaged economies. As this happened, large amounts of carbon dioxide were emitted into the atmosphere. This of course should have increased warming, but this is not what was observed. Instead, the temperature decreased. But why?

At this time, a significant number of aerosols were also emitted, from both industrial processes and volcanic activity. As mentioned in chapter one, aerosols reflect heat and light from the Sun back into space causing a cooling effect. They only remain in the atmosphere for a few years, however. In the mid twentieth century, so many aerosols were being released that they didn't just balance out the warming effect of the increased levels of carbon dioxide, they had a much greater effect, leading to an overall cooling[6].

As society learned more about the negative impact that aerosols have on the lungs, regulations came in to reduce their emissions. As this happened, the effect of the greenhouse gases then outweighed the effect of the aerosols, leading to a warming of the Earth.

The Earth was Warmer during the Medieval Warm Period

The Medieval Warm Period occurred from 950 CE to 1250 CE and experienced temperatures similar, or even slightly higher than they are today. So, if the Earth was warmer then, what do we have to worry about? Well, this Euro-centric view is not all it seems. It was actually only within the North Atlantic region that these temperatures were experienced[7]. When we look at global temperatures at that time, many regions were notably colder than today. There are theories as why the North Atlantic experienced this warm period, such as changes in sea currents which brought warmer waters north, but the point is that the average temperature of the Earth was indeed lower than it is now.

This Medieval Warm Period also coincided with a time of prosperity in Europe. This is used as an argument suggesting that an increase in temperature is actually beneficial. This may be true for some regions as they will have, for example, longer growing seasons. This, however, ignores the negatives in other regions around the world. It's as if some people only consider certain regions of the world to be of value.

Besides this, it is important to remember that the atmosphere now contains far more energy than it did previously. As mentioned in Chapter One, this is due to the increased levels of greenhouse gases. This extra energy has not only increased the average global temperature, but has led to observed extra heat waves, floods, droughts, cyclones, and sea level rise, that are having such devastating effects across the world. Not only that, but the scientific models tell us that, if we carry on as we are, then global temperatures, and their effects, will increase way beyond anything that the Earth has experienced for a long, long time.

It is also worth considering that whether the Medieval Warm Period was warmer than it is now or not is not the point. If natural causes had caused a warmer temperature in the past that in no way precludes the current Global Warming being caused by human activities.

Greenland used to be Green

Vikings settled in Greenland around one thousand years ago, whilst the North Atlantic was experiencing the aforementioned Warm Medieval Period. Temperatures in Greenland at that time were actually very similar to what they are now. There are reports that Greenland was green at this time, and not covered in ice. The Greenland ice sheet is at least four hundred thousand years old, however[8]. So, Greenland was certainly not green a thousand years ago. There is evidence of forests in southern Greenland, but these are at least four hundred thousand years old too.

Of course, some may wonder why it is called Greenland at all. Well, would seem that this was a thousand-year-old marketing strategy. The Norsemen who first settled there wanted to encourage others to do the same and so gave it a name that would suggest a green and vibrant land.

Antarctic Sea Ice is increasing

The amount of sea ice around the Antarctic increased from 1978, when sea ice extent records began, to 2016. This sea ice then reduced in both 2017 and 2018 before increasing again in 2019 and 2020. Surely if the sea ice is increasing, then global temperatures can't be?

Well, we saw in Chapter One, that the GRACE satellites show that Antarctica has actually been losing ice mass. Observations clearly show that whilst the area covered by Antarctic sea ice is increasing, the total mass of ice at Antarctica is actually decreasing.

Even if this wasn't the case, then why just pick out on area of the world that doesn't match the rest?

The increase in the extent of sea ice at the Antarctic is in sharp contrast to what is happening in the Arctic. Indeed, the decrease in sea ice in the Arctic far outweighs the increase in the Antarctic. Therefore, the total amount of global sea ice is decreasing[9].

Changes in the Sun's Output

The Sun is a remarkably consistent star. It does, however, have small natural variations in its output due to sunspots. These are dark spots on the Sun's surface, caused by strong magnetic fields. The more sunspots, the more active the Sun's surface is, and so more total energy, particularly in the form of ultraviolet radiation, is received by Earth.

Sunspots on the Sun's surface vary over a cycle of around eleven years. During this cycle, there is an overall variation in energy output of about zero-point-one percent, though this percentage is greater with the likes ultraviolet. Overall, this can cause temporary increases in temperature and rainfall in some regions, whilst decreasing them in others. Therefore, sunspots can cause slight variations in the global picture but this all part of an eleven-year cycle[10], not the increase over the last two hundred years that we are experiencing now.

Sunspot Activity has Increased since the Maunder Minimum

As mentioned above, sunspot activity varies with a cycle of around eleven years. There can, however, also be longer variations.

From 1645 to 1715 there was no sunspot activity[11]. This led to reduced temperatures on Earth, particularly in the Northern Hemisphere. The time is known as the Maunder Minimum and shows that extended periods of high or low sunspot activity can cause significant changes in temperature, even though variations in the standard eleven-year cycle do not have much effect.

Some studies over the last few hundred years seemed to suggest that the number of sunspots had been steadily increasing up to a so-called Modern Grand Maximum. So, could this increase in sunspots be what is causing the Earth to warm up today?

There had always been a good deal of debate about these numbers, as two different methods of counting sunspots gave different results. They can't both have been correct, so something was wrong. More detailed studies have now identified an error which has led to a correction in these numbers[12]. The evidence now tells us that sunspot activity has actually remained relatively stable over this time, so sunspot activity is definitely not to blame.

Changes in Volcanic Activity

Volcanoes emit carbon dioxide into the atmosphere. In the distant past volcanoes were so active that their emissions led to large increases in temperature, but volcanic activity has since decreased, meaning that this is no longer the case.

The amount of carbon dioxide emitted by human activities is between sixty and ninety times more than those emitted by volcanoes, from both eruptions and underground magma, each year. True enough, large eruptions take place every twenty years or so that match human emissions for several hours, but over the whole year the effect of this is still much smaller than that caused by human activities. Indeed, since these eruptions also emit ash and aerosol particles into the atmosphere, which reflect energy coming directly from the Sun, an increase in volcanic activity can reduce temperatures in the short term[13]. Therefore, volcanic activity is not the cause.

Cyclical Changes in the Earth's Orbit

As the Earth orbits the Sun there are slight variations in the orbit due to how the Earth interacts with other bodies in the solar system. Overall, the Earth's orbit is very circular but over time it does become slightly more elliptical, before becoming more circular again. This means that there are sometimes variations in the distance between the Earth and the Sun. This then changes the amount of energy that the Earth receives, and so affects its temperature. Added to this, the angle at which the Earth itself is tilted can vary, reducing or decreasing variations between seasons.

The overall variations take place with a period on the scale of around one hundred thousand years and is known as the Milankovitch cycle[14]. If the amount of land and sea were spread evenly across Earth, then this wouldn't cause much change in the average temperature over the year since the average distance to the Sun is still the same. At times it would be closer, but at other times it would be further away. There is relatively more land in the northern hemisphere, however. Since land is easier to warm up and cool down than water, variations within the cycle do cause the average temperature of the Earth to change.

Right now, we are in a time when the Earth's orbit is very circular, and the temperature changes are currently taking place over a far smaller time than those which could occur from changes in the Earth's orbit. The Earth's tilt, and orbit, are therefore not the cause of the current Climate Change.

Movement of the Sun within our Galaxy

The Sun is part of the Milky Way galaxy which contains hundreds of billions of stars. The Milky Way has a spiral shape and contains so-called spiral arms, in which the star density is relatively high, as well as areas outside these arms where it is relatively low.

Just as the Earth orbits around the Sun, our solar system also orbits around the centre of our galaxy. As it does so it moves in and out of these spiral arms. When we are within spiral arms, the Earth is closer to more stars and so is bombarded with more cosmic rays, i.e. particles that have been emitted from these stars. Some of these particles can make it more likely for types of low-lying clouds to form in our atmosphere. The effect of these extra clouds is that less heat from the Sun may reach the Earth's surface thus cooling the Earth.

As our solar system moves out of a spiral arm, fewer cosmic rays enter the Earth's atmosphere. This leads to fewer low-lying clouds, allowing more energy from the Sun to reach the Earth's surface and warm it.

This cycle of moving into and out of spiral arms is on the order of hundreds of millions of years and so may only have an effect over very long time period[15]. Therefore, it cannot be what is causing the current change.

Variations in the Earth's Magnetic Field

The Earth is constantly being bombarded by particles from the Sun. These would be potentially very damaging for life on Earth but fortunately the Earth has a magnetic field around it that deflects the majority of these particles, keeping us safe.

The strength of this magnetic field varies over time, meaning that the number of these particles reaching the Earth also varies. These incoming particles, like the cosmic rays mentioned above, can affect cloud formation and thus temperature. There is however no evidence that the natural variations within the Earth's magnetic field, for example it reversing (on average) every three hundred thousand years or so, has a significant effect on our climate. Currently the Earth's magnetic field is reducing in strength but has been doing so at a fairly constant rate for thousands of years rather than just over the last couple of hundred years[16].

It is therefore not what is responsible for our changing climate.

El Niño and La Niña

El Niño and La Niña are the warmer and colder natural fluctuations in the Pacific region around the equator. El Niño has warmer than usual water, leading to an increase in temperature and rainfall. La Niña is the opposite. They typically last between nine to twelve months though can last longer and occurs every two to seven years. Since there is fluctuation between El Niño and La Niña they cause no overall warming[17].

Variations in Global Temperature are Part of Natural Cycles

This is of course true; we know the Earth's average global temperature has changed many times in the past. There is no dispute on this. One thing we need to consider however is not just changes in temperature but the rate at which this change is taking place.

Generally, the rate of increase or decrease in temperature in the past has been at a much lower rate than that which is happening today[18]. When change occurs at a low rate, many species are able to adapt, and there are few extinctions. But what happens at a high rate of change of temperature?

Evidence tells that when rapid changes have occurred in the past, mass extinctions occur. Examples include those four-hundred-and-forty-four million years ago, two-hundred-and-fifty-two million years ago and two-hundred-and-one million years ago. This also makes sense. Rapid changes in temperature mean that species do not have time to adapt, whether in location, behaviour or through evolution, thus leading to extinction.

The last time the Earth was as warm as it is now was during the last interglacial, i.e. between ice ages, around one hundred and fifteen thousand years ago. At this time, whilst some areas of the earth, such as the polar regions, were two to three degrees Celsius higher than today, other regions were cooler, giving an average global temperature very similar to today's[19]. Even if it had been warmer, since all changes occurred at a significantly slower rate than today, life would have had a chance to adapt to many of the changes.

An increase of one point two degrees Celsius increase over two hundred years may not sound like a lot but compared to geological time frames this is high. Too high. Extinction is a natural event and has happened throughout history, but today's extinction rate is around one thousand times the natural rate. It seems that we are living within the so-called sixth mass extinction, and we are causing it.

Carbon Dioxide is a Consequence, not a Cause of Warming

If we look at the evidence on carbon dioxide and temperatures in the past, then an increasing concentration of carbon dioxide seems to lag behind increases in average global temperature[20]. This suggests that a temperature rise leads to having extra carbon dioxide, rather than the other way around. Maybe this empirical evidence then destroys our point about increased concentrations of greenhouse gases causing Global Warming. If this is right, then human activity is not to blame. We would then have to relook at the science. That would be ok since that's what our understanding of science does; it adapts as we gain more evidence, even when we think it was well understood. But is it the whole picture? Let's look at this in more detail to determine what is going on.

As mentioned previously, variations in the Earth's movement, the so-called Milankovitch cycle, take place over around one hundred thousand years. These variations mean that at points within this cycle, parts of the Earth are heated more or less than they would be if the Earth's movement was constant. The variations in temperature caused by this alone are small, however; they are too small to cause the magnitude of some of the changes in temperature which the records show. There must have been something else going on as well.

During periods when the cycle causes an increase in temperature, permafrost thaws releasing carbon dioxide and methane into the atmosphere. Organic matter then decays at a greater rate releasing even more carbon. At higher temperatures the oceans can't hold as much carbon dioxide either, meaning that more is released into the atmosphere.

Overall, this means that carbon dioxide concentrations do indeed increase when the average global temperature increases. But remember that the variations in the Earth's movements can't explain the larger changes in temperature which have been seen. So, what caused such large fluctuations in temperature? Well as mentioned above, the science is well understood on how greenhouse gases effect global temperatures. The fact that the temperature led to more greenhouse gases in the atmosphere, caused the global temperature to become significantly higher than it would otherwise have been.

So, is carbon dioxide a consequence of warming? Yes, but it is also a cause. Throughout history carbon dioxide has acted to amplify the increases in temperature due to the variations in the Earth's movement. Rather than an agreement against human caused Climate Change, this should actually be a cause of extra concern. It suggests that as the global temperature continues to increase, natural phenomena will lead to the release of even more carbon dioxide than we are emitting ourselves, causing further increases in temperature.

Much Higher Concentrations of Carbon Dioxide in the Past

Five hundred million years ago the concentration of carbon dioxide in the atmosphere was up to four thousand ppm, in other words it was around ten times higher than it is today. Surely if our understanding of the science is right then this should have led to a runaway increase in temperature?

Well, whilst in the short-term the Sun's output is relatively constant, it has changed significantly during its existence. It may be fairly constant now, but five hundred million years ago the Sun was significantly dimmer than it is today[21]. This combination of a dimmer Sun and very high carbon dioxide concentrations still led to a significantly warmer Earth. It is estimated to have been around seven degrees Celsius warmer than it is today. Indeed, if it were not for a high concentration of carbon dioxide then the Earth would have iced over. And for a time, it did. But surely the carbon dioxide concentration was too high for this to happen, even with a dimmer Sun. Is there something we are missing?

Well, we know that the concentration of carbon dioxide has naturally varied with time in the past. These variations in the past have been caused by volcanic activity, changes in plant growth and changes in ice cover. As it varied at this time, at times it would have been low enough to allow the Earth to freeze over.

So, the concentration of carbon dioxide on Earth has indeed, varied throughout its four and a half billion year 'lifetime'. The current rate of increase of carbon dioxide, however, is one hundred times that of any other within at least the last eight hundred thousand years. The concentration of carbon dioxide itself hasn't been this high for at least three million years.

Extreme Weather Events have Always Taken Place

One of the consequences of Climate Change as mentioned in chapter one is the increase in extreme weather events, such as floods, drought, heat waves and hurricanes. These have always existed. The difference is the frequency at which they occur.

Past extreme weather events cannot be used as a way to deny Climate Change. Science predicts that increases in temperature increase the likelihood of these extreme events occurring. This is matched by the evidence as the number of these events increase across the globe. It is true that some regions may suffer fewer extreme events but, looking at the big picture, the frequency of such events is increasing globally[22]. Not only that, but it is predicted to increase further as the average global temperature continues to rise.

Sea Level has been Rising for Thousands of Years

The sea level has varied greatly in the past. One hundred and twenty-five thousand years ago it was around five metres higher than it is now, whereas twenty thousand years ago it was one hundred and twenty metres lower. Since then, the sea level has increased to current levels. It seems sensible to say then that since the sea level has been rising over this time then the current rise in sea level is just part of that and nothing to do with Climate Change.

If we look at sea level over time in more detail, then we see a flaw in this argument. Over the last three thousand years the sea levels have gone up and down slightly, but there has been no overall rise or fall in this time. This is in marked contrast to the fact that sea level has increased by over twenty centimetres in the last hundred years[23].

Positive feedbacks would have led to Runaway Warming in the Past

This seems like a good argument. At times in the past, such as five hundred million years ago, there were higher global temperatures and higher concentrations of carbon dioxide. Modern science tells us of positive feedbacks such as increased water vapour, reduced albedo and release of carbon under melting permafrost. So why hasn't this led to runaway warming?

First of all, let's clarify that climate scientists aren't saying that there will be a runaway greenhouse effect on Earth. This is something that some in the media have greatly exaggerated.

Positive feedbacks exist, and these have caused a greater increase in global temperature than there would otherwise have been. So why haven't they led to runaway warming in the past? The simple answer is that there are, as mentioned above, negative feedbacks too. In certain scenarios the positive feedbacks outweigh the negative feedback, whereas in others the negative feedbacks outweigh the positive.

In general, the science says that the positive feedbacks give diminishing returns with increases in temperature. In other words, the greater the increase in temperature, the less the positive feedbacks further increase this temperature.

For example, positive feedbacks such as the reduction in albedo can't continue beyond the point that all the ice has melted, so have a natural cut off point. On top of this, negative feedbacks such as the increase in energy emitted from the Earth's surface as its temperature increases, and extra absorption of carbon dioxide due to increased photosynthesis in new areas will eventually cause the average global temperature to decrease[24]. But how much damage will have occurred by then?

Look at the Facts

An interesting tactic employed by those who deny Climate Change is to tell others to look at the facts. This is a way to suggest that they are the ones that know, hiding behind the truth that they are, whether consciously or subconsciously, cherry-picking evidence to suit their message. I wish it wasn't true, but the scientific evidence is overwhelming. Not only does the empirical data show that we are in a Climate Emergency, but we understand the science behind it. The science is clear on the reason; the increase in greenhouse gases due to human activities.

It's Young People's Fault

As someone who works with young people, I see far more willingness, on average, for them recognise the significance of Climate Change and to make changes to their lives than in older generations. There are of course many exceptions to this within both the young and the older.

This is a concern to those who deny Climate Change, meaning that it is important to attack the actions of young people. We hear that young people demand more electricity, more travel, more technology, more of everything, than any generation before them. This is done as a conscious attempt to deny young people the right to voice their concerns on Climate Change.

For some young people this may be true, relative to how we were when that age, but we are all products of our environment. Do not blame young people for being bombarded by messages from birth so that large corporations will make more profit. Also, the evidence shows that energy consumption increases, in general, with age[25]. So, it is not the young who are currently doing the most damage environmentally. It is just a myth that they are.

But let us not have one generation blame another. Let us all work together on this. Let's all look at the science and reflect upon what we want to be different for ourselves and for future generations. Then let us all, together, make positive changes that will make a genuine difference.

Does this including striking… yes, if that's what it takes for others to take notice. Individual action can and does make a difference but there must be a coming together of individuals to work differently. We must learn from each other, gain a sense that we are all in this together, and put pressure on corporations and governments who are doing the most damage so that they will take the required action too. We will look at this further in Chapter Eight.

Humans aren't Significant Enough, and it is Arrogant to Think that we are

Legend has it that King Canute tried to turn back the incoming tide. We ourselves are as powerless as he to turn back the water, turn off a hurricane or stop a drought. If we can't change such things, then how can we possibly think that we have the power to change the entire climate?

The simple reason is that there is clear evidence that we have had a huge impact on various parts on the climate before we even look at the impacts of increasing greenhouse gases.

For example, humans and their livestock now account for ninety-six percent of mammalian biomass[26]. This leaves just four percent for all wild mammalian biomass. Deforestation has led to huge changes to soil composition and quality, as well as to local rainfall. Human activity also released CFCs into the atmosphere which damaged the ozone layer. Thankfully we learnt from the science on this one, and the situation has improved significantly.

We also know that human activities have caused the concentration of carbon dioxide in the atmosphere increase from two hundred and eighty ppm to four hundred and fifteen ppm in the last two hundred years.

As human beings we have reached a stage where we do impact the whole world around us. How we affect it is up to us. Do we want to create a world of increased extinctions, heat waves, flooding and starvation, or do we want a better world for all?

Summary

Just because human caused Climate Change is real and understood by climate scientists, doesn't mean that everyone is able to accept this. It can be frustrating when someone sees something different to ourselves when shown the same evidence, but we do not go into this as blank slates. Our past experiences and the information we have been bombarded with, shape how we view the world. Whilst this has led to division and anger for some, with the right change it can also bring us together. We will look at what we can do both as part of a group, and as individuals, in Chapter Nine. For now, try not to lay blame, be angry with or belittle those who see things differently. Work together. There can be no them and us. We are all part of the same family. Remember, this is our climate, our future, and there is a better way.

CHAPTER 3

ENERGY TRUTHS

We rely upon natural energy resources for so much in our lives. Without them our lives would be very different. Indeed, we would have no electricity, no cars, no planes, no heating, no cooking, no refrigeration, and we would be without a great many of the materials that we rely so heavily upon. Energy resources are therefore fundamental to modern life, and whilst use has been made of some of them for thousands of years, it's never been to the extent that we do today. Our modern lives very much depend upon them, but they come at a cost.

Let's start this topic with a few truths which are not always said;

1. There is no such thing as a completely clean or green energy resource, renewable or not
2. Some energy resources, however, are far, far better than others

3. Without the burning of fossil fuels, we cannot currently provide sufficient energy to meet global demand
4. We cannot afford, environmentally, to continue to burn fossil fuels
5. Therefore, our demands must change, our lives must change

The fact that no energy resource causes zero environmental damage may come as a surprise to some. This is especially true since many energy companies publicly proclaim theirs to be zero-carbon. But they are not telling the whole story. The issue is that the stated electricity generation emissions are based purely upon what is emitted at the point of generation. This means that any carbon emissions, or other environmental issues, from the likes of transport, mining and construction, all of which occur before the actual generation of the electricity, are not considered.

Within this understanding, however, we need to recognise that there are huge differences in the environmental impacts of each. Let us now then, consider each energy resource in turn, in order to recognise the pros and cons of them all, in order that we can make informed decisions for a better future.

Fossil Fuels

Fossil fuels have a very bad name environmentally, and for good reason. They are also, by far, the most used energy resource on Earth, providing over eighty percent of global energy[1]. It is sensible then to now look at why they are so bad for the environment, along with why they are currently used so much.

The three main fossil fuels are coal, oil and natural gas. Fossil fuels are formed over many millions of years from dead plants and animals, and contain the carbon and energy that was once within these organisms. In order to harness this stored energy, they are burned. It is this burning that releases carbon dioxide. Indeed, this process has contributed more to Climate Change than any other activity[2].

The carbon stored within the fossils fuels was once in the form of carbon dioxide in the atmosphere. Through photosynthesis within plants, this carbon was absorbed from the air, becoming the carbon which make up the plants, then later the animals which eat them. Since it was once in the air, then what is the harm in putting it back? Well, the issue is that this carbon dioxide was removed from the atmosphere a long, long time ago, and over a very lengthy time period. Whereas now, it is being emitted, not only millions of years later, but over a very short time frame.

It is the burning of fossil fuels then, that has not only made the biggest difference to way we live today, but also accounts for the biggest rise in the concentration of greenhouse gases, which we find in our atmosphere today.

It is also worth noting that not all fossil fuels are equal. When we compare the amount of carbon dioxide released by each of these, we find that burning coal releases significantly more carbon dioxide per unit of energy, than the other two, with oil being next, and with natural gas releasing the least amount of these three[3].

The picture is not this simple, however. We are all aware of oil spills that have caused enormous amounts of environmental and commercial damage. For example, BP's 2010 oil spill in the Gulf of Mexico released around eight-hundred million litres of oil into the ocean[4]. This human-caused disaster, directly killed eleven people, led to the death of thousands of sea birds, and had a devastating impact on marine life, including turtles, fish, and coral. It also put around ten thousand people temporarily out of work.

Natural gas is often considered to be the 'cleanest' of the fossil fuels, but this only looks at the actual burning of the fuel. Whilst it is a mixture of different gases, the vast majority of it is methane, a gas we discussed in Chapter One. It has been observed time and time again that natural gas fields leak some of this methane into the atmosphere. This makes them far more damaging that they would be otherwise. It is estimated that natural gas could be even worse than coal if methane leaks are greater than two-point-seven percent of the total[5]. When we consider the fact that some of these fields have been found to emit notably more than this[6], then we should recognise the need to look at natural gas in a new light.

And when it comes to fracking, in which high pressure liquid is forced into rocks underground, causing the rock to crack, releasing oil and natural gas, the amount of leaked methane is frankly terrifying[7]. All this evidence clearly tells us that, whilst natural gas is often marketed as the clean fossil fuel, it is nothing of the sort. What we should, therefore, be striving for, is the reduction, and then the complete halt, in extraction and burning of each of these three fossil fuels.

The fossil fuel industry is also far more dangerous to work for than any of the other energy resources. Worker deaths from accidents per unit of energy are far, far greater with fossil fuels, than any of the others[8]. For example, both natural gas and oil are often transported along vast pipelines. From the 1950s to the modern day, pipelines have exploded, not only causing significant emissions, but destroying homes and the local environment. In many cases direct death also occurs, including one-hundred-and-thirty-seven people dying from a such an explosion in Mexico, in 2019[9].

The burning of fossil fuels emits, not just carbon dioxide, but also carbon monoxide, nitrogen oxides, sulphur dioxide and particulate matter such as soot. These cause huge amounts of damage to the health of living things and are estimated to cause millions of deaths per year. One study even showed that up to one in five deaths every year is caused by the pollution from fossil fuels[10]! This includes the increased risk of children dying from leukemia[11]! Surely this is something that none of us want to continue.

Added to this, we should also recognise the fact that we only have a finite amount of fossil fuels on Earth. They are running out. If we continue to burn fossil fuels at current rates then it is estimated that we have around thirty years left of oil, forty years left of natural gas and seventy years left of coal[12]. Estimations do vary since energy demands may fluctuate in the future and new reserves may be found, but these give us an idea of the kind of timeline we are talking about. What this is telling us is that, even if the burning of fossil fuels wasn't a major cause of Climate Change, then we would still have to be looking at how to reduce and move away from their use. We cannot wait that long, however. To do so would put so much life on our planet in jeopardy. A priority for all needs to be as speedy a phase out of their use as possible.

With so many related issues, why are they used to the extent they are? Well, fossil fuels first became a significant energy resource for human use during the eighteenth century. They are easy to obtain, store and transport, burn well, release lots of energy and have traditionally been relatively cheap. All this led to a lifestyle of far greater convenience than in the past. The vast majority our major industries, transport and electricity depend upon fossil fuels.

Fossil fuels are reasonably cheap for the consumer too. Many of us assume that they are just cheap because they are. We rarely question it since we then don't have to pay so much money to fill up our cars, run the heating, or use the electricity in our homes. But is it that simple? The fossil fuel industry is the most profitable in the world. So, what makes them so cheap and yet the profits so high?

When we look into this in more detail, we find that a major factor as to why fossil fuels are so cheap for the consumer is because they benefit from huge subsidies. The International Monetary Fund (IMF) estimates that in 2015 alone the fossil fuel industry received more than over five trillion dollars in direct and indirect subsidies[13]. To put this in context, this is more than ten million dollars per minute and between six and seven percent of the Gross World Product. In other words, the subsidies, note, just these subsidies, for fossil fuels alone, amount to around six-point-five percent of the monetary value of all finished goods and services across the world! This doesn't even include the trillions of gallons of water for cooling that the fossil fuel industry receives for free every year.

But where do these subsidies come from? Well, they come from you, the taxpayers. Your tax money is being used directly to support the fossil fuel industry, and to make a lot of money for a relatively small number of people. It is no secret that fossil fuel companies provide significant amounts of money and support to political parties, individual politicians and their campaigns[14]. Politicians need funding for their campaigns so welcome the financial support, so then support the companies in return. For the burning of fossil fuels to stop, they must be taken out of politics. Subsidies do make fossil fuels cheaper for the consumers, but when we consider that many other energy resources are already as cheap as fossil fuels, and that environmentally we cannot go on releasing carbon dioxide into our atmosphere, then this practice must stop as soon as possible. It is also worth considering here, that if fossil fuels are only viable when they receive huge government subsidies, then are they really viable at all?

We should also add here that, as seen in the UK in 2022, a society currently reliant upon fossil fuels, is susceptible to huge spikes in the cost of living. No, we cannot carry on are we are.

Within all of this, it is important to make it clear that fossil fuel companies, and the people who own them and work for them, aren't setting out to damage the environment; it's just that their profits come first. They do, however, spend huge amounts of money trying to convince us that they do care, all whilst extracting and burning more and more, for greater and greater profit. Since they have made such vast profits, and caused so much damage to the environment, should it not be these companies which finance the required actions, and initiatives, to reverse the environmental damage that they have done?

In short, we have a lot to thank fossil fuels for, but their time has come. We must embrace other energy resources, whether for electricity, transport or other needs. The problem is, we can only transition away from fossil fuels if we have suitable alternatives to transition to. Let's then have a look at these now.

Nuclear

Nuclear power stations use uranium or plutonium to release the energy required for electricity. This happens through a process called nuclear fission. In this method a large atomic nucleus is split up into two smaller nuclei, releasing energy. Since no actual burning is involved, no carbon dioxide is emitted. Nuclear power stations are also reliable and can generate vast amounts of electricity, so seem like the perfect solution. But are they really? Let's look deeper.

The main issues involving nuclear power stations stem from the fact that the waste from the nuclear fission process is radioactive. A typical nuclear power stations produces over twenty tonnes of highly radioactive waste from fuel every year[15]! These radioactive materials can cause significant damage to health, including an increased probability of certain cancers, and so need to be disposed of safely. What's more, some of these materials retain a dangerous level of radioactivity for many thousands of years.

There are three main types of radioactive waste produced, depending on their properties. Low level nuclear waste is stored within metal containers and are kept within concrete vaults. Intermediate level waste is stored, with concrete, inside steel drums. These are then placed within underground facilities. High level waste is generally liquid and is mixed with molten glass in a process known as vitrification. This is then placed within steel containers, and they are kept in a cooled store for around fifty years, before moving them to an underground facility. This all sounds ok, but the containers are not even designed to last more than a few hundred years. Given that the combination of glass and steel may well corrode over a time period way smaller even than this, then how is that going to stop radioactive waste leaks in the future? Is this really what we want to be leaving for future generations?

There is also the significant issue regarding nuclear accidents. Nuclear power stations for electricity have only existed since 1954. In that time there have been many such accidents, with the four largest being Windscale (now Sellafield) in 1957[16], Three Mile Island in 1979[17], Chernobyl in 1986[18] and Fukushima in 2011[19]. Each of these led to the release of significant amounts of radioactive material into the environment. We can keep saying that designs have improved, and they are now safe, but the evidence tells us that this is not something we can know with absolute certainty. Furthermore, nuclear power stations are generally built near the sea to make use of the water for cooling. What will happen as sea levels rise and the sea water gets into the stations?

Another issue with nuclear is the expense. Whilst the cost of using renewable energy has reduced over time, the cost of a nuclear power station is getting more expensive[20]. The main expenditures involved include the initial set up as well as it's later decommissioning. This is the shutdown of the power station, which includes all the aforementioned issues in dealing with the nuclear waste. Like fossil fuels, large subsidies are also given to the nuclear industry. If it were not for these, then the electricity from nuclear power stations would be extremely expensive.

Despite this there is an argument that nuclear power is necessary, in the short to mid-term at least. This is because renewables may well not expand fast enough to deliver the cost-effective, reliable electricity that we are now used to. Added to this, many more people across the world are rightly gaining access to electricity, in order to improve their lives. One suggested proposal to cope with a reduction is fossil fuels, includes the building of over one hundred nuclear power stations every year for the next thirty-five years. When we consider the evidence on the construction of these however, then this looks unlikely to be feasible. Whilst we are told that modern nuclear power stations will take just a few years to build, when we look the construction of such stations, we find that many are years behind schedule, and billions of pounds over budget.

Saying that, despite the carbon emissions due to the construction of the buildings and the mining and transport of the uranium, nuclear power stations do have a significantly lower carbon footprint than fossil fuels. The fact that they can also run day and night means that they can fill a gap until the electricity from the more environmentally friendly renewable resources becomes more reliable and accessible. But given that they are more expensive, take longer to set up, and cause greater emissions than most renewables, and have the added risks regarding radioactive waste, how large a role should they really play?

There is another nuclear energy resource which is worthy of note. This is nuclear fusion, in which two small nuclei, such as from hydrogen, join together. Like fission, fusion releases energy without the need for burning. It is also the process which takes place within stars, causing them to emit huge amounts of heat and light.

The issue with fusion is that it can only occur at extremely high temperatures within a confined space. This means that large amounts of energy need to be put into it. More energy is released than is put in, but the longest anyone has been able to maintain the fusion reaction, thus far, is only five seconds[21].

Whilst far more research needs to be carried out, its potential is enormous. For starters, Earth contains huge amounts of hydrogen within water. Fusion also creates no radioactive waste and has no carbon emissions. If we could get it to work on a large scale, it would completely revolutionise our energy. No wonder scientists are working so hard to make this happen.

Solar

One common observation is that the Sun shines every day, and that this energy from the Sun is vital for our existence. Despite that, we currently make use of just a very small proportion of it, though that number is increasing. Solar energy can be obtained every day, has no fuel costs and is probably the most environmentally friendly of all energy resources.

There is a carbon footprint in first making, transporting and putting the solar cells together, but over their lifetime they release only around six percent of carbon dioxide compared to coal powered electricity, per unit of energy[22].

When first introduced, due to set up costs, solar energy was expensive compared to other energy resources. Over time, however, its price has dramatically reduced, and could well soon become the cheapest form of generating electricity. This reduction in price, alongside an increasing environmental awareness, means that the demand for solar energy is growing at an increasing rate.

Most solar energy makes use of solar, or photovoltaic, cells. These convert light directly into electricity. Solar farms can contain hundreds, thousands, or even millions of panels of these cells. Whilst they are able to generate a lot of electricity, they do take up land which could otherwise have other uses, such as plants. They can however also be set up in deserts, on land fill, or floating on reservoirs, where they also reduce evaporation. It is not just in solar farms where they are used either. Small scale solar is becoming more common on roofs and in remote locations which aren't connected to a larger grid. Barcelona even became the first location to install solar pavements. Solar cells also generate electricity at temperatures, and humidities, which may well cause alternatives to fail.

A less common use of solar energy is in Concentrated Solar Power (CSP). Here, it is the heat, rather than the light from the Sun, which is utilised. Curved mirrors concentrate the sun's rays, leading to temperatures high enough to heat water, produce steam, and turn a turbine. An added benefit of CSP is that it is easier, and cheaper, to store energy in the form of heat, rather than electricity. This means that the energy obtained can be stored to be used at a later time, when it is needed. For example, heat from the Sun's rays can be stored and then used at night-time, when the light-dependent solar cells are unable to work. Currently CSPs are relatively inefficient, and the mirrors make them expensive to build. The concentrated heat has also been known to kill bats and birds. So, whilst there are issues, they certainly have a good deal of potential moving forward.

The biggest issue, of course, with solar is that it only generates electricity during the day. Sunny days produce the most, but electricity is also generated on cloudy days since, as we all know, there is still light when it is cloudy, just less. Indeed, that can generate so much electricity during the day that they can drive down energy prices significantly.

At night, however, no electricity form solar is generated. Also, as noted above, different amounts of electricity will be generated on different days. This all leads to the idea that solar cannot be relied upon. But how unreliable is it really? We know that the Sun will shine every day, even if varying degrees get through to the solar cells. A great deal of work, as we will see later, is being done on energy storage. It is already in place in many areas of the world but needs to increase. As this improves, then the main issue regarding solar will reduce dramatically.

An added bonus with solar is that no one can monopolise, or control the amount of solar energy we receive from the Sun. This means that, as long as the solar cells are in place, prices cannot be manipulated in order to create more profit.

From all of this, it should be clear that the pros that solar brings far outweigh the cons. Its use will play a major role in the fight against Climate Change. Just imagine if all those subsidies going to the fossil fuel industry went to investing in solar and energy storage.

Wind

As the air is heated by different amounts, these differences force air to move along, in other words create wind. The energy in this wind can be used to turn turbines which can then lead to the generation of electricity.

Whilst there are costs, and carbon emissions, in the construction and maintenance of the wind turbines, once they are set up there is no direct fuel used and overall wind energy is one of the lowest carbon emitters[21]. Another advantage of wind is that it uses around ninety-nine percent less water than generating electricity from fossil fuels.

Like solar, the energy from wind was originally expensive, but these days wind turbines produce electricity at the lowest cost of all new energy ventures. One of the reasons why costs are reducing so much is because many turbines are now built at high elevations. This means that longer blades can be used. These places also, often, have more wind.

Wind turbines can be set up on the seas or on land. Offshore wind turbines benefit from typically stronger winds and don't use up land space. They are generally built far from shore so they can't be seen and reduce complaints. This does increase the length of cables which are required for the electricity to travel along, however. Also, due to the sea being such a hostile environment for such things, their lifetimes tend to be shorter. They also have increased installation and maintenance costs, making them more expensive than their onshore counterparts.

Onshore wind farms are easier to assemble and maintain. Whilst they do use up land, they generally take up no more than one percent of the land upon which they sit. This land then can be used for other uses, including agriculture, recreation and conservation. A NIMBY 'not-in-my-back-yard' attitude, often citing noise or the way the turbines look, creates obstacles to their construction. The wind turbines can also damage migrating birds and bats, but newer turbines are designed to reduce this by using slower turning blades as well as avoiding migration paths.

Another cited issue with wind turbines is from a misreading of a study leading to the suggestion that wind turbines are causing Climate Change. I know! In the study it was found that the land local to wind turbines could increase in temperature. But if we think about the science then it should be clear that wind turbines do not release extra energy, they just make use of energy that is already in the air. What is actually happening is that at night-time the rotating turbines draw down warmer air from above. This means that the land below the turbines is warmer at night, but other places will then get cooler. There is no net effect on the average global temperature. Indeed, using wind turbines rather than, for example, fossil fuels, means that the amount of global heating will be a lot less than it would have been otherwise.

As well as the large-scale wind farms, there are also smaller micro wind turbines. These are typically singular and are used to pump water, charge batteries and to provide electricity. They are particularly useful in remote, rural locations, that otherwise wouldn't receive such energy. These smaller turbines are far less prone to complaints about spoiling the views of landscapes and are much quieter. Some micro wind turbines are also combined with solar cells, boosting their energy capacity. Cost remains an issue with these micro turbines, but economy of scale, and enhanced technology, should help reduce these costs for the future.

The best-known issue with wind energy of course, is that some places have far more wind than others and that the wind doesn't always blow. This means that the turbines will not always be turning, and so electricity will not always be generated. To overcome these natural variations in wind, the electricity can be fed into a larger, e.g. national, grid, so that whilst it may not be windy in some places it may well be windy in others. Energy storage systems also help here.

The low carbon aspect of wind energy means that it has to play a part in our energy sources of the future. Its use is increasing and will continue to do so into the future.

Geothermal

Below the Earth's surface there is a significant amount of heat energy that can be utilised. Whilst this is most obvious when volcanic eruptions take place, this subterranean heat can be made use of all over the world, in the form of geothermal energy. Note that it's not just in volcanic area in which this can happen. One study, for example, suggests that warm water from Cornwall's granite crust could provide a fifth of the UK's power needs[23].

Most geothermal stations make cold water flow under the surface. This is then heated by the rocks below. This heated water is then pumped back up to the surface where the energy can be made use of. This process can happen all day every day, which can help compensate for the natural variations in the likes of solar and wind.

Like the other renewable sources above, no fuel is burned, so beyond the construction and maintenance there are few carbon emissions. Having said that, when the water comes back up to the surface it can contain dissolved gases such as carbon dioxide, mercury, arsenic, and boric acid. Carbon emissions, however, are still less than ten percent per unit of energy than those of a coal power station[24]. It is also cheaper.

There are issues which include finding suitable sites, the possibility of soil subsidence, and that the stations can be both noisy and smelly. Current estimates suggest that we are currently only making use of a very small percent of suitable geothermal energy, however, in the future it could produce most, if not all of our electricity. The likes of solar and wind are better environmentally, but it is good to know that there is the potential to have such a significant, reliable energy resource moving forward.

Waves and Tidal

These two energy resources make use of the energy within the seas. Waves are caused by the wind and the tides are caused by the gravitational pull of the seas from the Moon, and to a lesser extent the Sun.

Like solar and wind, there is no fuel requirement, so nothing is burned. This means that the only carbon emissions, and cost, come from the construction and maintenance of the wave and tidal turbines.

One issue with wave energy is that some days the waves transfer more energy than others. Tidal energy, however, is far more predictable. Though the tides only come in or out at certain times within the day, we can accurately predict when, and by how much. We should also note that the number of sites where electricity can be generated from waves, and to greater extent tidal, is fairly limited. They will never be able to be our main global energy resource, but they can certainly play a part.

One issue which should be mentioned with tidal is that they may slow down ocean currents. This would reduce the vital exchange of water, and the nutrients it carries, both from place to place, and from the bottom to the surface. Having said that, if Climate Change continues to increase as it has been, then these currents will be affected to a far higher degree, as the ocean temperature and salinity alter[25]. It may well be that the positive contributions for tidal will greatly outweigh the negative.

Right now, wave and tidal energies are more expensive than other renewable sources, but with the furthering of technology and an increased economy of scale the costs will reduce. There will surely be a use as we transition to a world without fossil fuels.

Hydroelectric

Hydroelectric power stations harness the energy from gravity. Water from high up moves downwards, turning a turbine, and generating electricity. Since it is the flow of water, there are no fuels burned and so carbon emissions just come from the construction and maintenance. Or so we thought.

Hydroelectric power stations include the building of a dam to block the river. The water behind the dam then fills a large reservoir, and this water can then be released when electricity is required. With large power stations these reservoirs can lead to the flooding of huge areas of land. For example, the Three Gorges Dam in China led to over one million people losing their homes and being displaced to other areas[26]. Habitats were also destroyed, and a significant amount of wildlife was lost. This process can also impact the movement of sediment, as well as the migration of fish. The displacement of people leads to the need for new houses, factories, schools, etc. All of these adding to the carbon footprint of the dam.

But we have still not mentioned the largest emission of greenhouse gases from hydroelectric. In large, still, bodies of water, plants will break down and decay. This decay is often without sufficient oxygen and so rather than releasing the carbon dioxide that the plants absorbed as they grew, it releases methane. And as we saw in Chapter One, methane is an extremely potent greenhouse gas.

This means that hydroelectric power stations are nowhere near as environmentally friendly as they are sometimes portrayed to be. Yes, they are renewable, and no fossil fuels are burned, but they are a long way from being the answer to our climate problems.

A big advantage that hydroelectric has over several the other renewables, is that the water can be released, and so generate electricity, at any time. In fact, hydroelectric can respond to variations in electricity needs much faster than any fossil fuel. The larger stations are also able to generate vast amounts of electricity. Another big advantage of such power stations is that they can store energy. When there is a surplus, energy can be used to pump water back up to the reservoir, for use at a later time. How much these pros outweigh the cons is a matter of some debate.

There is, also, an alternative hydro power. Smaller 'in-stream' turbines are placed within a river or stream. These make use of the movement energy of the water, without the need for a reservoir. Therefore, they work in a similar way to wind turbines, which make use of the movement energy of air. The moving blades of the turbine may of course affect local water life, meaning that they need to be designed to reduce any negative impact.

As always, great care must be taken with these. Flooding may occur where turbines are placed too close together and some have diverted paths of streams, but with careful design and management there is a good deal of potential in these systems. These do generate far less electricity and can only be used at places with flowing water, but, where suitable, can be used in remote locations which are not connected to large electric grids.

In the short to medium term hydroelectric will play a vital role as we move away from fossil fuels. Over time though, as other renewable energy sources increase, and energy storage improves, hydroelectric will eventually become obsolete.

Biomass

As mentioned previously, the energy in fossil fuels originally comes from that stored by plants in photosynthesis. Biomass works the same way but instead of that energy, and carbon, being absorbed millions of years ago, this happened far more recently. The plants, or their by-products, are then burned. Whilst this burning does release carbon dioxide, the timeframe over which this all happens is far shorter. Also, since this burning can't release more carbon dioxide than the plant absorbed in the first place, the biomass seems to be carbon neutral. But is this the whole picture?

Biomass is indeed marketed as being carbon neutral, with premise that the balance between the amount of carbon absorbed and released is maintained. In order for this to occur, it is said that any plants which are grown and burned will then need to be replenished at an equal rate. Where trees and other plants are removed at a faster rate than they can be grown, deforestation occurs. This also leads to less plant life on Earth, meaning less carbon dioxide can be absorbed. Still, this seems very doable.

When we look further into it, however, we find that there is a lot more to it than it first seems. This is because the biology of a forest is not so simple. For example, removing trees also reduces the ability to store nutrients and water within the ground. This leads to a reduction in soil fertility, negatively affecting the ability of more plants to grow there in the future, thus reducing the long-term carbon storage of trees[27]. Also, as discussed further in Chapter Four, the soil itself stores huge amounts of carbon. The reduction in its water and nutrients causes the soil to emit carbon into the atmosphere. And then there are all the other species that rely upon the trees. When the trees are cut, many other species will lose their homes, their source of food and water, and their protection. In other words, local biodiversity will dramatically reduce, again leading to an overall loss in carbon storage. No matter how well the trees grow back, it will take a long, long time for all of this to recover.

This effect can be reduced, but not stopped, when the biomass comes from so-called sustainably grown plants. It can also come from waste products, such as from saw and paper mills, as well as discarded plant parts from food crops. We know though, that some of this plant waste should remain on the fields to support soil health. Another short-term source of biomass comes from burning invasive species which are damaging local wildlife and so it is right that they are removed.

Another issue is that, whilst industry propaganda insists that biomass comes from waste and the like, the evidence tells is that some of these companies, which make the wood pellets, are actually removing logs from whole trees, straight into their mills[28]. In 2019, two countries in the Baltics even exported over three million tonnes of wood pellets. This equates to the loss of at least two hundred square kilometres of forest in just one year[29]. With so much biodiversity loss in the world, as further discussed in Chapter Seven, how can this be ok?

Even where the biomass comes from supposedly sustainable sources, there are other factors to consider. These include how much water and fertilizer is required to grow the plants as quickly as they are required. Suitable choices of plant materials need to be made, including those which grow quickly. Those plants which grow back after being cut back close to the ground are also, generally, preferable. Indeed, many trees are actually not suitable for biomass since they can take many decades to grow back. It is also vital to avoid the growing of monocultures, in which large numbers of the same species are all planted in one place. Nature does not work like this for a reason. The lack of biodiversity they produce essentially leads to a reduction in life; the very thing we are trying to avoid.

Another issue with biomass is that the burning of it releases other substances, along with the carbon dioxide. Indeed, such burning, as seen in the likes of wood burners in homes, is the largest source of small particle air pollution in the UK[30]. To give an idea of scale, this produces three times as much as all road traffic! This already sounds bad enough, but then we learn that wood stoves are responsible for almost half of the cancer-causing chemicals with the air[31]! This is information that everyone with, or considering buying a wood stove, should know.

As well as wood, there are many other different sources of biomass. These include crops, algae, and waste. There is also a fairly recent development that uses grass[32]. The grass is cut several times every year. It is then put into a large chamber, where it breaks down over time, releasing methane. This can then be used in the same way as natural gas. Now, we have already seen the affect that methane has on our climate, so it is vital that no leaks occur. Whilst grass is a new source of energy, we know that plenty of methane escapes within the fossil fuel industry, so a good deal of research and development will be needed to ensure the same doesn't happen here. It is also still on a relatively small scale, and the effects on local biodiversity need to be investigated further, but there is the potential for some use.

Overall, we should be able to see that biomass is not the carbon neutral source it is portrayed to be, but it is certainly better than fossil fuels. One big advantage that biomass has over many of the other renewables, is that it can be used to produce electricity on demand. This means that as the world moves away from fossil fuels, biomass will surely be needed, for a while at least. But is the burning of trees and other plants really the solution to saving the planet?

Ground Source Heating

Ground Source Heating literally removes heat from the ground. In other words, it is a form of geothermal energy. This removed energy can be used for heating the home and for hot water.

It works as a heat pump, where a loop of pipe is buried in the ground, such as under a garden. A fluid passes through the pipe and gets heated from the ground. A compressor then increases the temperature further. This compressor does need electricity to work but, overall, the carbon emissions are much lower than many other heating systems, particularly where the electricity comes from non-fossil fuel sources.

Ground Source Heating can be very expensive to install but the reduced running costs mean that over time it does become financially suitable.

Solid Cooking Fuels

In many parts of the world, materials such as wood, charcoal, coal, animal dung and crop residue are burned for cooking. These emit damaging air pollution, often within poorly ventilated homes. This leads to a negative impact on the lungs, hearts and eyes of those who inhale it; often women and young children. Indeed, this household air pollution is a leading cause of early death, so needs to change.

Improved, more efficient cooking stoves can make a huge difference in these areas. They require less fuel, meaning fewer emissions, less disease, and fewer deaths. Also, the reduction in fuel means that far less has to be gathered, which reduces safety issues in the gathering. These cooking stoves can be relatively expensive, however, and financial support, as well as education, is often required to ensure their implementation. Supporting such projects can have a real positive impact on the lives of others.

Burning Waste

A move towards zero waste is vital as part of reducing our environmental impact. The current situation, however, is that we are continuing to produce vast amounts of it. Whilst we still have such waste, it can be burned, releasing energy which can be used to generate electricity.

The burning of waste releases carbon dioxide, increasing its concentration in the atmosphere. This is obviously the opposite of what we want, but it is better than burning fossil fuels. In addition to this, if organic waste ends up in landfill instead, then much of it will decay anaerobically, that is, without oxygen. This causes the production and release of methane; a far more harmful greenhouse gas than carbon dioxide.

This burning is also, often, offered as a means to reduce waste, but let's make something clear; burning waste does not get rid of it, it merely changes from one form into another. The volume of solid waste is reduced which means less goes to landfill, but some is also emitted as gas. This includes such substances as lead, mercury, cadmium, nitrous oxide, sulfur and hydrochloric acids, fluorides and small particulate matter which can damage the lungs. Without burning, much of this would remain as part of the solid waste rather than being released. In some places, waste power plants can remove these pollutants, meaning that they will not end up in the atmosphere or the water. Where this does not happen, however, they will be released.

We must also consider the issue when the burning of waste is cheaper than the alternatives. This may mean then that some will be burned rather than treated in better ways. It can also mean that reducing waste is not encouraged, though we know it must happen. In the end our 'waste' must become so valuable that the last thing you would ever want to do is to get rid of it.

Hydrogen

Hydrogen is essentially a way to store energy. Hydrogen is a flammable substance which can be gathered to be burned later, releasing energy. There are no direct carbon emissions and the only by-product is water. Despite hydrogen being by far the most common element in the Universe, finding it pure on Earth is extremely rare. This is just as well, given that it is flammable! There is, however, plenty of hydrogen on Earth as part of compounds, such as water.

There are three main ways in which to obtain the hydrogen. These are known as grey, blue and green hydrogen, and vary greatly in the environmental impact from their production; not all hydrogen is created equally.

Grey hydrogen is currently the most common form of creating hydrogen. Here, fossil fuels in the form of natural gas or coal are used. Natural gas is combined with heated steam at high pressure, causing hydrogen to be released. With coal, it is heated with limited oxygen. This creates carbon monoxide, which is reacted with steam. This leads to the creation of both carbon dioxide, which is released into the atmosphere, and hydrogen. This hydrogen from coal is also sometimes called brown hydrogen. We have seen above that our reliance upon fossil fuels is not sustainable and we cannot consider grey, or brown, hydrogen, as being a positive solution.

Blue hydrogen again comes from the use of fossil fuels. The difference now is that the emitted carbon dioxide is captured and stored, well, some of it is. We will look at Carbon Capture and Storage in more detail in Chapter Six. The capture that does occur, makes blue hydrogen more environmentally friendly than the grey, but it is not the solution either.

With green hydrogen, rather than using fossil fuels, water is split up into hydrogen and water by a process known as electrolysis. The only increase in greenhouse emissions comes from this use of electricity. When low carbon sources, such as solar and wind, are used, then green hydrogen is effectively a low carbon energy source.

In whatever way the hydrogen is produced, when it is burned, it produces water, which is a greenhouse gas. When this happens with green hydrogen, this is no more water than was lost in obtaining the hydrogen in the first place. Also, by mixing hydrogen with natural gas in the correct proportions, the amount of carbon emissions is reduced without affecting the efficiency of the fuel. Obviously, we need to move away from all use of natural gas, along with the other fossil fuels, but this would reduce emissions during the transition period.

For transport, the hydrogen is stored in hydrogen fuel cells. This fuel can then be used generate an electric current, which runs the motor. These cells are relatively small in size for the amount of energy they can release. Having said that, in most counties there is very little infrastructure for hydrogen cars, despite the benefits. They are also expensive at the moment but will become more affordable over time. Hydrogen cars may well be part of our future and have many benefits over electric cars in terms of fuelling and range, but hydrogen is a flammable gas and related accidents at hydrogen producing plants have led, understandably, to a greater emphasis on electric.

Now, it does require more energy to extract the hydrogen than can later be obtained from it. This means that hydrogen is not an energy resource as such but is certainly a useful store. Another benefit of hydrogen is that it is also already widely used within the chemical industry. This means that a lot of technology is already designed to make use of it.

So, hydrogen will surely place an important role moving forward, but we need to ensure that this hydrogen is green, not grey, brown, blue or other.

Storing Energy

We have seen that some of the most environmentally friendly energy resources cannot provide energy all the time. This means that energy storage will play a crucial role in the future. This means that, in times of high generation, lots of energy can be stored. It can then later be released to cope with the needs of the people all day, every day.

The energy can be stored in several different ways. Whichever method is used, it both needs to store sufficient amounts, as well as be able to react to increased demands in electricity in a very short time. We have seen that hydrogen can be used to store energy, but let's look at the other options.

One way is to store this energy up high. In pumped hydro systems, excess electricity is used to pump water from a low reservoir to a high reservoir. This water can then later be released downwards, turning a turbine and generating electricity when it is needed. Whilst this has its place it is obviously limited by suitability of locations as dictated by changes in height, available water, as well as space for reservoirs. This is currently by far the most used energy storage system.

An alternative is to move railcars to a greater height, rather than water. A motor and a generator are essentially the same thing but work in reverse. These railcars each contain a number of these, which act as motor on the way up and generators on the way down. A big advantage that these have over water, is that they can remain at height for a long time without losing energy, whereas a reservoir will lose water due to evaporation.

Another way to store energy is as heat. An example of this is in the aforementioned concentrated solar power plants. These use energy from the Sun to heat molten salt. This salt then remains at a high temperature for many hours, allowing the energy originally from solar to be used for hours after sunset.

A further method of storing energy is within flywheels. These consist of a large rotating cylinder which is made to rotate, thus storing the energy. These are designed to have very low friction, and are kept within a vacuum, meaning they can store energy for extended periods. When required, the flywheel can then be connected to a generator in order to produce electricity. These flywheels are durable and so last a long time, allowing for regular 'charge' and 'discharge' as required.

The other main way to store energy is as electricity. This is currently more difficult to do than in the forms mentioned above. The electricity can be stored in either a battery or a capacitor. The technology on these is improving all the time, with them becoming cheaper, safer, and more reliable. Their extended use will play a significant role moving forward.

The Grid System

As we move, as we must do, away from fossil fuels and to a greater reliance on renewables, there will be more variation in how much electricity can be generated at any one time. This needs to be taken account of within the whole setup. There is no one solution for all, however. Different scenarios, in different locations, will need a variety of approaches.

In some areas the use of large national or even international grids would be the best plan. In this case, when it is not sunny or windy in one area it may well be in another. Therefore, the total electricity entering the grid won't vary that much. The main issue with this system though is that the further electricity travels along cables, the more energy is lost to the surroundings.

In other areas a microgrid would be more beneficial. This reduces the distance that electricity has to travel, but relies upon only local energy being made use of. This micro-grid could even be linked to, and so feed into, or from, a larger grid, as required.

Microgrids are also very useful in remote rural areas. This is particularly true in regions which contain some of the one billion people on Earth who do not currently have access to electricity. Instead of the high costs of linking these remote areas to large grids, the local, natural resources can be used.

Summary

For so long we have relied upon fossil fuels, but the times they are a changing. As should be clear from the above, our dependence upon fossil fuels needs to end. The faster we make that change the better. Much of the energy sector recognises this need and is already making positive changes. The percentage of electricity generated using other energy resources is increasing year by year, and electric vehicles are continually being developed.

We must remember that there is no one size that fits all. A combination of different resources must be used, in conjunction, to provide us with the required energy needs. A mixture of solar, wind and geothermal, with other renewables supporting, looks to be the best solution. There will be the need for so-called 'bridging' energy resources, such as biomass, hydroelectric and nuclear, in the short-term, but eventually we will also want to move away from these too.

The argument that fossil fuels are cheaper than renewable energies is no longer true. Many of the renewables are now noticeably cheaper than fossil fuels and are continuing to reduce in price. Also, whilst the likes of biomass and hydroelectric may not be ideal, they show that we can also have electricity on demand. With improvements in energy storage, then solar and wind will become more reliable too.

Moving to low carbon energy resources is not only necessary environmentally, meaning essential for life, but is often already cheaper than fossil fuels. So why isn't more being done? Sadly, the political motivation is still behind, in part at least, due to huge lobbying from the fossil fuel industry. This will not change until enough people refuse to vote for those who aren't doing enough to make it happen. We look into this in more detail in Chapter Eight.

The biggest message here should be that our energy needs must change. Not only must we switch away from fossil fuels, and to higher efficiency appliances, but our behaviour itself must change. No energy resource is zero carbon. The idea that by switching to renewables, as we must, will mean that we can continue to use as much as we like, also needs to change.

For now, ask yourself, do we really have the right to use any energy we can pay for, if it is causing such damage to our world? In the future, with proper support, the less polluting energy resources will be able to meet all our energy demands. Until then, we reduce. There are a number of ways to help do this discussed in Chapter Nine. Is this really too much of a hardship compared to the consequences of Climate Change? Remember, this is our climate, our future, and there is a better way.

CHAPTER 4

FOOD, GLORIOUS FOOD

Food is a basic requirement, an essential human right. None of us can survive without sustenance. Food growers and providers across the world, whether self, community, global, small or large, do a vital job in keeping us alive. Agriculture today provides for seven point nine billion people[1], seventy billion farm animals[2], and many millions of pets. Since these numbers are so vast, it is surely worth looking at the environmental impact of providing this much food and, if required, how we can improve it.

A look at the large-scale research tells us that food production accounts for twenty-six percent of the world's greenhouse gas emissions[3]. On top of that, half of the world's habitable land is used for agriculture[4]. Seventy percent of freshwater taken from rivers and lakes is used for agriculture[5]. And when we look at the impact on biodiversity, we find that there is fifteen times more mammal livestock biomass on Earth compared to wild mammals[6].

This does not make for pleasant reading, but maybe that's the way it must be in order to feed such numbers, or is it? We owe so much to agriculture and farmers, but times are changing, as the Environmental Crisis becomes clearer and clearer. We must then look at whether we are able to make positive changes with the types of food that we eat, the agricultural practices used, as well as the transport, waste and packaging of food.

Types of Food

As we look at the different types of food, we must recognise each of the different environmental issues involved. Along with greenhouse gas emissions, we should also consider land use, water use, other pollutions, and biodiversity. Let's then look at how different food types affect each of these in turn.

Different Types of Food and Greenhouse Gases

As mentioned above, over a quarter of greenhouse gas emissions comes from our food. But are all foods equal in this regard? There are a variety of ways of comparing the impact of different foods. When comparing them, we can relate the amount of greenhouse gases emitted to calories, to protein, or to a variety of other key nutrients.

A sensible place to start is to look at calories. The only caveat on this is that we will only compare foods that have a reasonable quantity of calories. In other words, we will compare foods that people eat for the energy they bring, rather than very low-calorie foods that people eat for flavour or various nutrients.

When we look at the evidence on this it becomes clear that there is a vast difference in greenhouse gas emissions per calorie. For example, per calorie, the production of beef results in thirty times the emissions from beans[7]. Let's just clarify what that means. If we replace beef with the same number of calories of beans on our plate, then we reduce the greenhouse emissions due to that part of our meal by thirty times. That's huge!

Of the foods that we eat for calories, beef is the largest, per calorie, contributor to greenhouse emissions, but when we look at other foods there is a very clear trend. Animal-based foods have far greater emissions than those which are plant-based. In fact, the only plant-based foods which are exceptions to this are very low-calorie foods, which we eat for the nutrients they contain, such as tomatoes. Since they are so low calorie, a lot of tomatoes need to be grown to make the same number of calories. For such foods this is not a fair measure.

After beef, the highest emission foods, in order, are farmed prawns, lamb, farmed fish, cheese, poultry, milk and then pork[7]. These are clearly all animal products. Note: coffee and chocolate are also high up on the list but coffee, at least, is not particularly consumed for the calories. There are definite issues with chocolate, however, primarily linked to the cutting down of forests to grow plantations, along with the fact that it takes a long time for cocoa trees to grow suitable amounts[8].

At this stage some may say that just looking at emissions per calorie is not fair. There is a myth that animal products are needed for proteins, so let's now look at the emissions per one hundred grams of protein. Again, it is only fair here to compare foods which we eat for protein. It would not be a fair comparison to compare high protein with low protein foods here.

Let's again compare beef to beans, both of which are eaten, in part at least, for their protein. We find that per one hundred grams of protein, beef produces twenty-five times the emissions of beans[7]. Again, let's clarify what that means. If we replace beef with the same amount of protein from soybeans on our plate, then we reduce the greenhouse emissions due to that part of our meal by twenty-five times. Again, this is huge!

Here we have a simple change which we could make that would reduce the greenhouse gas emissions of that part of our life by twenty-five to thirty times. If we understand the effect of greenhouse gases on our planet, and we care about not just other life on our planet, not only our own friends and families, but also ourselves, then surely this is something that, both as individuals, and as society, we need to consider very carefully.

When we compare all the different high protein foods, we again find that it is the animal-products which emit the most greenhouse gases per one hundred grams of protein. Other examples include eggs releasing sixteen times more than nuts, and poultry emitting almost thirteen times more than peas[7].

At this stage some may point out that beans such as soybeans are actually part of the problem, rather than the solution. It is well known that in certain parts of the world, deforestation takes place in order to grow soybeans, as well as for the likes of beef farming. Indeed, in some regions a considerable percentage of the deforestation is, indeed, for soy. So surely soy is the issue rather than meat? To understand this further, let's have a look at what then happens to it. When we look at the numbers, we see that seventy-seven percent of this soy is fed to livestock[9]. Seventy-seven percent! In other words, the majority of soy is grown solely to feed cows, chickens, pigs, and farmed fish. Thirteen percent is used to make soybean oil, which is a common cooking oil. Another four percent is used in industry. This leaves just over six percent of soy being produced for human food. Given this evidence, it is clearly not the human consumption of soybean foods that should get the blame for the destruction of the rainforests.

By this stage it should be obvious that animal products contribute to the emission of far more greenhouse gases than plant-based products, but not all animals are raised in the same way. Surely, if we just use so-called sustainable animal product producers, then we can continue to consume animal products without the high greenhouse gas emissions. Let's have a look.

Firstly, for every food source there is a variation with the amount of greenhouse gas emissions. This can be due to a variety of factors including location, intensity of farming, food supply for the animals, whether deforestation has occurred for the land, as well as a number of farming techniques. Whilst poultry and pigs are generally produced intensively around the world, beef shows greater variation. This partly depends upon whether the cows are also used for dairy products or not, since with dairy cows the total emissions are shared with the dairy products, making the emissions for beef alone, lower. It can also depend upon the terrain, temperature and the fertility of the soil. Land management practices, including avoiding de-forestation, also have a big impact.

What is clear however, it that even the lowest emission animal-based products still produce more emissions than the highest emission plant-based products. For example, when we look at the protein comparison, soybeans, peas and nuts always have lower emissions than meat and dairy.

A major factor as to why beef and lamb lead to such a high emission of greenhouse gases is due to the fact that cows and sheep are ruminants. Their digestion process leads to the emission of methane, which we know to be a very potent greenhouse gas. Just how much this occurs does depend upon their diet, but methane is always released. One example is when seaweed is added into their diet. This is found to reduce the amount of methane they emit dramatically[10]. It may be that on a global scale the quantities of seaweed required would make this impractical, but research continues into how this can be improved.

We should note though, that even if these animals emitted no methane at all, then their total greenhouse gas emissions would still be much higher than those from plant-based products. The non-methane emissions in animal agriculture come from many sources, including converting peat to agricultural land, growing animal feed, fertilisers, irrigation, the slaughter process, transport, machinery and waste.

The vast majority of plant-based products emit no methane. The main exception is rice. Rice is a staple for almost half of the world's population and provides around twenty percent of calories consumed worldwide. Rice is typically grown in waterlogged soils in paddy fields. Since there is little oxygen here, the microbes in these soils emit methane. This is, of course, factored in when we look at the emissions per calorie from rice, and it is found that the total emissions are still far lower than any of the animal-based products, but this is still something we want to reduce. There are techniques which can be used, such as the System of Rice Intensification[11]. This not only reduces the emission of methane but also increase yields, reduces water use, requires fewer seeds, less fertiliser, and seems to make the rice more resistant to drought, flooding and storms. On the other hand, it is more labour intensive, initially at least, but it is a system that is getting more popular with rice growers around the world.

Let us also remember that even if the best producers of animal-based proteins didn't emit that much more than the worst producers of plant-based proteins, there still isn't much of an argument to eat meat and dairy. How would it be fair to compare best with worst? If we can achieve best practice with one, then why not with the other. Indeed, even if all meat and dairy were produced at the lower end, which it sadly is not, then plant-based products would still be better in terms of emissions.

So, purely in terms of greenhouse gas emissions, reducing meat intake and switching to a more plant-based diet is clearly the best option. Meat intake should be at least reduced, and the meat which is consumed should be predominantly poultry and pork, rather than beef and lamb. Also, with every single type of food, switching to a more sustainably sourced product is preferred. If we all moved away from the highest emission products, then the global emissions of greenhouse gases would decrease noticeably.

As part of this we must also accept that not everywhere in the world has access to all the different plant-based products. Maintaining some meat and dairy may well be required, in relatively small amounts, in parts of the world. But that is not an issue in the materially wealthier countries, where plant-based products are becoming ever more accessible and popular.

Different Types of Food and Land Use

As mentioned previously, half of the world's habitable land is used for agriculture[4]. When considering this, please remember that all of that land was once wild. Of course, we need to feed the world's population, but are there ways of doing so without using as much land so that it can have other uses, such as rewilding?

A fair way to compare foods on this, would be on the amount of land use per calorie. As we saw with greenhouse gas emissions, when we look at the evidence, the trend is clear. Animal-based products use far more land area per calorie than plant-based ones. Beef herds, again the worst, require over ninety times more land per calorie than beans[12]. This means that if we switch from beef to beans, then for the same number of calories, we could reduce the agricultural land which is currently used for beef herds by a factor of ninety! Think about how much land could then be rewilded, how much deforestation could be reduced, stopped, and even reversed. Lamb is again the second worst, with, in order, cheese, beef from dairy herds, milk, pig meat, farmed fish and eggs being the foods which require the most land per calorie[12]. These are all animal products.

If, as before, we compare land use per one hundred grams of protein instead, for high protein foods, we get a similar picture[13]. This time lamb is the worst, requiring eighty-four times more land for this protein than beans. Beef, cheese, milk and pig meat also require significantly more land for this protein than any plant-based product.

Now of course, we could compare a high protein animal product to one of the plant-based products which hardly contain any protein and argue that the plant-based version is far worse, but that wouldn't be comparing like with like. No one eats lettuce for protein.

Using agricultural methods such as factory farms, rather than grass fed animals, does reduce the land use, but there are other significant other issues with their use. There are some people who, very understandably, want all livestock to be grass fed. In order to do this, and meet current demand on animal products, then the land use required would be enormous. Indeed, the only way in which we could have all livestock reared on grass, would be to significantly reduce the amount of meat that we consume across the world.

In total, animal-based products account for just eighteen percent of the world's land calories, thirty-seven percent of the protein, but seventy-seven percent of the agricultural land use[14]. This, of course, includes land which is used for grazing or growing crops purely for animal consumption. The evidence is clear; if we want to reduce how much land is used for agriculture, and rewild that land so that wild plants and animals can thrive, then we must reduce our consumption of animal products.

As time has gone on, the amount of land used for agriculture has increased dramatically. This is partly from an increased demand for food, and partly due to the type of food we are consuming. Whatever the reasons, and we look into some of these in more detail in Chapter Eight, we need to reduce the amount of land used for agriculture. For now, only by changing the types of food that we are eating, can we possibly hope to have enough land to feed us all, as well as protect life on Earth.

This change, and increase, in land use for agriculture, is arguably the largest impact that humans have ever had on the environment. When we also consider that farming techniques have also increased yields, then we can see that the current situation would, have otherwise, been far worse. With proper support, as we adjust the food types we consume, land will be allowed to go back to its natural habitat; becoming wild once more.

Faced with this information, many producers of animal-products, such as beef farmers, may argue that grazing lands are significant carbon sinks, and so capture and contain large amounts of carbon. This is indeed true. It doesn't, however, consider the whole picture. Reducing the farming of animal products decreases how much land is needed to feed the human population, allowing that land to be used for rewilding and forests, and thus acting as an even larger carbon sink.

Other land issues linked to agriculture are desertification, over-farming of land leading to low soil fertility, soil erosion and depleted ground water. This forsaken land must be treated so that it can once again act as a carbon sink. This requires investment and incentives, whether the land will then be used for rewilding, or for farming. Making the land productive once more takes time, effort and money, but can we really afford not to?

The alternative is to obtain ever more land for farming. This is often done by deforestation, which clearly has a huge impact on both emissions, and biodiversity. Therefore, all farming should only be on that land which is already developed, and not on, or even next to, virgin land, such as forests.

Overall, the evidence tells us that in order to reduce the amount of land used for agriculture then we should move away from animal-based products and towards plant-based. This would allow both the reduction of land for agriculture and so more land for rewilding, but at the same time, the ability to grow more food for in increasing human population.

Different Types of Food and Water Use

We know that changes in climate are leading, and will further lead, to how much precipitation is received in different places across the world. Some areas will receive too much, whereas others will get too little. Other areas rely significantly upon runoff from glaciers. As these glaciers reduce, potentially even disappearing entirely, these regions will receive far less water. In other words, water will become an ever more precious commodity. Therefore, it is vital that we protect our water supplies. Any way in which we can reduce our water use, must be worth investigating.

Also, so much water is currently used for irrigation, that rivers, streams, and ground water are often depleted. This can also lead to significant disputes when water is removed up-river, leaving little water for those who depend upon it further down. Indeed, many rivers have dried up entirely as a result of human activity.

We find that seventy percent of freshwater use is for agriculture[15]. The remaining thirty percent is used for industry and households. Often when there is water shortage, particularly in certain parts of the world, households have to reduce their use, though agriculture will often still receive significant amounts of it. This seems reasonable, as long as people have enough water to drink, cook and wash, since we need to have enough food to eat. But are there ways to reduce the amount of water that agriculture requires?

Therefore, let's look at how much water is required for different types of food, and see if there is anything that we can do to grow enough food whilst protecting our waterways.

Initially we will, like before, compare this using calories. As before, let's do this for foods which are eaten for calories rather than other nutrients. It may now not come as a surprise that animal products, once again, come up as the highest use of water per calorie[16]. For example, farmed prawns require about five times more water than nuts. Except for foods such as tomatoes, that we don't eat for the calorific content, all the foods with the highest water use are animal products. Other plant-based products, such as vegetables and cereals, use far less water. If we instead compare the foods in terms of protein, for high protein foods, cheese comes up as requiring the most water. Nuts are also high up, but the rest of the high-water use foods are from animals.

The evidence tells us that to protect rivers, streams, and ground water around the world we need to reduce the amount of water we are removing from them. Since agriculture is responsible for the majority of water use and, in general, animal-based products use significantly more water than plant-based products, a move towards plant-based would help protect the Earth's fresh water, whilst also providing nutrition for us all.

Different Types of Food and Other Pollution

As well as the emission of carbon dioxide and methane, agriculture is responsible for many other pollutions leading to issues such as eutrophication. This is where the amount of nutrients, such as nitrogen, is increased so much within a body of water, that an excessive amount of algae is produced. As these algae die, they decompose, reducing the concentration of dissolved oxygen in the water. Fish and other water creatures can then no longer breathe, and so die off, creating dead zones.

Nitrogen is necessary for plant growth, and a key part of many fertilisers. The problem comes when more nitrogen is added than is necessary for the plants. If too much nitrogen is taken up by the plants, then they won't grow as well. If too much is left within the soil, then it can end up in ground water, rivers and streams.

Excess nitrogen also emits nitrous oxide, which is an extremely potent greenhouse gas. It is therefore extremely important to have the correct amount of nitrogen within the soil. This is not as easy as you might think. Farmers generally know their soil well but getting the right amount of fertiliser can be complicated. Different plants require different amounts, and even the same plant needs different amounts at different times during its growth. Farmers want to use the correct amount, but the information on how much to use often comes from companies that make the fertilisers. The upshot of this is that, to ensure there is enough, the soil is often given too much.

Agriculture leads to seventy-eight percent of all eutrophication[16], but maybe there is a way to reduce it? When we compare different foods, we find that farmed prawns are the biggest contributors to eutrophication per calorie, followed by, in order, beef, farmed fish, pig meat, lamb, poultry, cheese and milk. Plant-based products lead to far less eutrophication. The eutrophication from animals comes, not only from fertilisers, but also by animal waste. This naturally ends up on the surface and this is washed into streams and rivers by rain.

Improved agricultural techniques and understanding are leading to less use of fertilisers in many places across the world, but a switch to more plant-based products will reduce this further. It should also be noted that there is growing evidence that rock-based and plant-based fertilisers, rather than the often-used animal-based, or artificial fertilisers, can produce yields which are at least as good[17].

Different Types of Food and Biodiversity

The advent of agriculture is probably the biggest change in human life, but its expansion has also been one of humanity's largest impacts on the environment. This change has transformed and destroyed habitats. Agriculture is then a leading factor in the extinction of species.

Only four percent of mammal biomass on Earth is from wild mammals. Hang on, let's say that again. Of all the mammals on Earth, only four percent of their biomass is from wild mammals! Thirty-six percent of the remaining mammal biomass is humans, which means that a massive sixty percent is from livestock[6]. We also find that seventy percent of birds are farmed poultry, leaving only thirty percent as wild birds[18].

This is not a situation we would want for our planet, as ever more species move into extinction. Indeed, many scientists state that we are currently experiencing the Earth's sixth mass extinction. Let us then consider why this is happening and what we can do about it.

Climate Change itself is a significant factor, as temperatures change, flooding and droughts increase, ice melts, sea level rises, and hurricanes increase. Other factors include land being taken from the wild, habitat destruction, deforestation, the implementation of biodiversity-poor monocultures, waste and other pollution.

Whilst we have seen many issues with beef, one argument that is sometimes given in support of cows is that, in certain parts of the world, large grazers such as bison and buffalo, lived and grazed in huge numbers in the past, leading to high biodiversity amongst other species too. Since they didn't create a negative impact then, how could cows now? This sounds like a good argument, until you look in more detail. In the wild, the large herbivores migrate over huge areas, moving from one place to another, to return the next year. Inside fenced land the cows are unable to do this. Even when the cows are grass fed, rather than confined within quarters, the same land is regularly grazed, robbing the land of its nutrients. This has happened with vast areas of land across the world. Grass fed livestock has many positives over that which is raised via intensive farming, but in order to mimic nature and the migrations that occur, whilst also providing for current meat consumption, a huge amount of land would be required. Indeed, for countries in the western world, the required land area would often be greater than all the land in that country! So, whilst natural herbivores on grasslands are an incredibly important part of the ecosystem, and their movement and waste support the soil, this is not what is happening in animal agriculture. It's as if nature got it right and humans got it wrong. It may well be that there is a place in our world for animal agriculture, but it needs to be at a significantly reduced amount, and we must learn from nature, in order to make it as sustainable as we can.

One product that has increased greatly in use, as has our understanding of its effects, is palm oil. The vast majority of this is planted on deforested land[19]. A walk through a palm oil plantation is eerily silent, as the insects, birds, and other animals that lived there previously, are no longer supported. Despite the very apt negative publicity on palm oil, it does use less land than other oils. The one thing we don't want to happen then is to replace it with those which use even more land, leading to even greater deforestation. This doesn't mean, however, that we should promote the use of palm oil. There is so-called sustainable palm oil, which is from companies which declare that the no more deforestation, treat their workers fairly, create wildlife zones, amongst other requirements. This sounds good but the likes of Greenpeace have deemed that the rules are not enforced and therefore not reliable[20]. Maybe we should be having a proper look at what type of product we are eating, rather than just replace one damaging ingredient for another.

All the above basically tells us that real change must occur…

One way to increase biodiversity is to change the food types we are eating and rewild land. It has to be done properly though. Rather than just growing the fastest growing trees, we must look to nature. Forests contain a large variety of native trees, which then leads to the creation of a huge abundance of biodiverse life. Thus, when rewilding we must mimic what nature does.

Not all soil is suitable for trees, and rewilding should suit the soil and climate. For example, in some areas, grasslands would be more suitable. Essentially, there is no one size that fits all. Wherever land is reclaimed for nature, local knowledge must be used, in order to create the best solution for that location.

Agriculture and biodiversity can also support each other[21]. By adding a variety of cover crops, which are plants which are primarily grown to improve the soil, its quality can be improved. This leads to an increased variety of nutrients and reduces the chances of infestation. Plants, including trees, can also be added to provide shade, increase biodiversity, and reduce seeds being blown away by the wind.

Another aspect of biodiversity we should look at is the bees. These are a vital part of the ecosystem and are well known for the vital pollination which they carry out. Indeed, without bees many crops and wild plants, would fail, leading to significant food supply issues. Evidence suggests, however, that bees are in decline[22]. This is due to habitat destruction, pesticides, and Climate Change. So, should we eat honey to support bee numbers?

Well firstly we should recognise that across the world there are around twenty thousand species of bee[23]. Of these, only seven are honeybees. Therefore, despite common misconceptions, the vast majority of bee species do not make honey. Honeybees will also compete with wild bees for nectar. This isn't a big issue where flowers are plentiful, but where they are not, then wild bees can be outcompeted. Not only is this bad for the wild bees themselves, but it's also not great for pollination. This is because whilst honeybees are very good at collecting nectar and pollen, but they then take much of it back to the hive, rather than to lots of other flowers. This means that they don't pollinate flowers as well as many of the wild bees. So, eating honey does support honeybees, but not the majority of bee species. In order to maintain, or even grow bee populations, we must support their habitats. Find out which native plants are best for wild bees, plant them where you can, and allow the plants on lawns, verges and the like to grow without constant mowing.

Healthy Soil

We all know that a key factor for life on our planet is soil. The soil stores huge amounts of carbon. In fact, it stores more carbon than all plant life and the atmosphere put together[24]. Having healthy soil also leads to healthy plants, which leads to healthy animals, and healthy us. It also has such a large impact on water, as well as carbon, that having healthy soil really does support a healthy climate. So, what actually is healthy soil?

Healthy soil contains plant matter and huge numbers of microbes. Plants taken carbon from the air through photosynthesis. Much of this carbon is directly to grow the plant, but a significant proportion goes down into the roots where it then feeds microbes. In return the microbes supply the plants with nutrients. These microbes then store huge amounts of carbon within the soil. Having so many microbes also supports the flow of water through the soil.

So, what's the problem? Firstly, ploughing or tilling soil is an agricultural process that has happened for thousands of years. This is done to break up the solid structure of the soil to aid drainage. It can also damage the roots of weeds, thus preventing them from competing with the desired crops. All good then, right? Well, this process also exposes more topsoil to the air. This means that the wind, and water, will remove this topsoil, causing soil erosion. This, in turn, reduces the fertility of the soil, meaning that less food or other plant life can grow successfully. It also increases sediment within streams and rivers, leading to a reduction of species in those areas too. In fact, over ploughing, or tilling, of the soil is considered to be one of the reasons why empires of the past collapsed. And it is certainly why we are seeing so many examples of enhanced soil erosion around the world today. One scary stat is that, since the 1970s, around one-third of the world's topsoil has been lost[25]! This is even more disturbing when we consider the fact that it takes hundreds of years for just a centimetre of topsoil to grow, but it can be lost within a year.

What's more, when too much soil erosion occurs and soil fertility reduces, farmers may be encouraged to add more and more chemicals, such as fertilisers and pesticides to the soil. This use of fertilisers and pesticides may well increase the yield, to start with at least. Since they are so successful, in the short term, this can encourage some farmers to use them more and more, especially given the large amount of information they are given from those companies selling the products.

Is this then the whole story? Sadly not. This use of high levels of chemicals is one of the reasons which led to the separation of crops and animals within farms. In other words, it caused them to move further away from what nature had been teaching us all along. It also led to specialisation, in which monocrops are planted, massively reducing biodiversity wherever it is done. It has also supported the growth of the larger plantations. This, in turn, allowed these larger farms to undercut the smaller ones, forcing them out of business. The larger organisations could then sweep in and purchase the smaller holdings at a reduced cost, further increasing their share of the market.

Ok, but what about the soil itself? It has been found that the use of chemical fertilisers, pesticides and herbicides just masks the problem of degrading soils. A key ingredient to any healthy soil is the microbes it contains. These microbes have a symbiotic relationship with the plants, as they support and help each other. The use of artificial chemicals, however, can actually kill these microbes. In other words, these added chemicals actually kill what healthy soil needs! Over time then, as the soil becomes less and less fertile, farmers may add more and more chemicals, until the soil is completely dead[26]. The microbes also help hold water and carbon and other nutrients within the soil. Their loss then makes the soil drier and causes it to release carbon dioxide into the atmosphere. Given how much carbon is stored within the soil, this is a real concern[27].

Not only this, but since plants grow better in soil with healthy microbes, clouds are more likely to form there, causing rain to fall on these areas. Conversely, as the microbes die and the soil loses fertility, the bare ground produced gets hotter in the day, causing warm air to rise and push rain clouds away. Therefore, the very micro-climate around the farm changes. When this is done worldwide, of course, the whole climate changes!

Ok, but is there any direct effect upon us? What we know is that these artificial chemicals enter the water cycle and indeed come into our drinking water. Our bodies contain huge numbers of microbes which are necessary to maintain a healthy life. These added chemicals, however, also kill these microbes and therefore damage our own health. Indeed, there are studies which link the use of artificial chemicals in agriculture to increases in certain types of cancer[28].

Not only that, but the death of soil leads to desertification. As this occurs there is a reduction in food security. Whilst it can happen anywhere, some areas are affected far more than others. People obviously cannot live where there is no food, and so this forces people from their homes; from the land their ancestors worked on for generations. This obviously increases poverty and environmental refugees[29].

An interesting side note here is that the many climate deniers are also those against immigration, despite the fact that their inaction on Climate Change is actually fuelling the immigration they oppose. Just imagine if we treated all human beings as one big family and stopped separating ourselves from each other!

So, ploughing and tilling are known to cause damage, but they still occur throughout the world. So, what is the alternative?

One way of farming successfully without ploughing or tilling is in direct drilling, in which seeds are drilled into unploughed soil. It has been shown that this method reduces soil erosion, supports soil biology, saves fuel from ploughing, leads to less herbicide and pesticide runoff, and keeps the carbon in the soil[30].

Another way is with so-called 'Regenerative farming' or 'Agroforestry'. Whilst the process varies from region to region, to fit in with the local climate and biodiversity, studies show that it not only increases the fertility of the soil, but it also leads to more food per area of land[31].

The key is to view the farm as a living ecosystem. As well as no ploughing or tilling, central to this is biodiversity. Different plants have different needs and bring a variety of benefits. Monocultures, in which every plant is identical, reduces soil biology. If we learn from nature and having a wide variety of plant species, then the soil is much healthier. The exact plants used as cover crops depend upon the local environment. It may be that trees, grasses, or other native species are most suitable.

As part of this, there should be living plants in the ground, rather than bare soil, at all times. The so-called cover crops hold in water when it rains. This in turn leads to the storage of a higher number of microbes within the soil, which leads to greater plant growth, and encourages extra local rainfall. This also leads to extra carbon being stored within the soil itself. Nutrients are also less likely to be lost when the soil has plants within it. In short, we need to learn from nature, and instead of fighting against the beauty of the natural world, work with it.

As mentioned previously, animal agriculture can cause measurable damage to our environment, but in learning from nature it is important to recognise that herbivores play a positive role too, but only if done in the right way. When animals eat plants, parts break off, adding carbon to the soil. Their hooves then push plant material into the soil and their excrement also provides nutrients and microbes. But all this gain is far outweighed by the negative if the livestock does not follow nature as a guide and follow a suitable migration. Any region of land should only have these animals on it for a relatively short time before moving on. This means that, whilst the total area used may be large, only a very small proportion of that land will have farmed animals upon it at any one time. This again tells us that, whilst there is a place for animal agriculture, it needs to be reduced significantly, as does our reliance upon animal products.

When considerable efforts are made, then previously degraded soil, including that deemed to no longer be suitable for plant life, can be restored. Some plants grow far better in damaged soils than others and over time, with careful management, and increasing biodiversity, the health of the soil will improve so it becomes suitable for agriculture or rewilding once more. Of course, having such healthy soil doesn't only bring environmental benefits, but also increases food security and supports the local economy.

Transport of Food and Drink

This one is obvious, isn't it? If we want to reduce the greenhouse gas emissions from our food and drink then we should just use local products, right? Less travel means less fossil fuels burned so less emission of carbon. It is so obvious that it often doesn't get questioned. It is, however, important to question all of our assumptions from time to time, so let's have a look.

To genuinely understand the impact of eating local, we should look at the proportion of greenhouse gas emissions which come from its transport, compared to other aspects of each food. When we do this, we find that this transport, with one notable exception, is a very small fraction of the total emissions due to that food[32].

This means that it is the type of food, rather than the distance that the food has to travel, which makes the biggest difference. For example, any beef, including that from a local farm, emits far greater emissions than any beans, including those from the other side of the world. Replacing meat and dairy with plant-based products has a far greater impact on reducing greenhouse gas emissions than buying local. It seems that the idea of eating locally to be more environmentally friendly is a myth.

It is true that eating local beef rather than beef raised on deforested land in, say, the Amazon region, would lead to a reduction in emissions, but if you really want to reduce the environmental impact of your food then should you be eating beef at all? Paradoxically, in some cases, eating locally may even cause an increase in emissions. This is because in most regions around the world, the local climate dictates when in the year is the best time to grow certain produce, and when it is not. These days, however, consumers want their favourite produce all year round. For a local farm to supply this out of season then either energy-intensive production methods, such as the use of hot houses, need to be implemented, or it needs to be refrigerated, or preserved in some other way. This can lead to greater carbon than importing from afar.

We mentioned above that there was one exception. Most food is transported either by land or sea. The exception to this is air-freighted, in other words, transported by plane. Only a small proportion of food is freighted in this way[33], but it can emit between fifty and one hundred times the amount of greenhouse gases than when shipped. Air-freighted foods should therefore be avoided, but which are they? The only foods which would ever be air-freighted are those from far away but with a short shelf life and so would perish on a longer ship journey. It is difficult to always know, since currently laws require no such labels on packaging. If unsure then assume that such foods have been air-freighted, until you find out differently.

Overall, we have seen that local food does not generally lead to fewer emissions. There may be other reasons for buying local, such as supporting the local community, but we shouldn't think that by doing so we are acting environmentally.

Food from our Oceans

Thus far, we have focussed on farmed foods, but many people also rely upon the oceans for a major part of their diet. With the oceans we are not cutting down forests, fertilisers and pesticides are not used, wild animals are not replaced with domestic ones, so surely this is much better? Let's have a look.

What the evidence tells us that the total numbers of fish, marine mammals, sea birds and sea reptiles, have halved in the last fifty years[34]. When it comes to the number of older, larger specimens of the different species then the numbers are much lower still. Almost ninety percent of large fish stocks have become disturbingly low[35]. So, why has this happened? Over-fishing is major factor here. This is where an unsustainably large number of fish are removed from our oceans. Therefore, by reducing the amount of fish that we eat, their numbers will be able to increase once more. There are, however, other factors to consider, including habitat damage, notably coral reefs and mangroves, coastal development, pollution, liquid chemicals and plastics, and Climate Change leading to both rising temperatures and acidity of the oceans.

Some countries are implementing initiatives to mitigate this. Certain fishing methods are becoming banned and limits on fish numbers are being created. Fishing grounds have been closed and illegal fishing is being cracked down on. Activities such as bottom trawling, in which weighted nets are dragged across the sea floor, must be stopped. Not only do they lead to a high proportion of bycatch, but they often occur around submarine mountains, or seamounts. These seamounts are hot spots of biodiversity, often containing deep sea corals. The bottom trawling destroys these ecosystems, and they could take hundreds of years to rebuild.

So, our consumption of fish is completely unsustainable. It is true that many coastal populations around the world rely upon fishing, not just for food but also for their livelihoods. It is often larger, international fishing vessels that are damaging their local fish stock. Only by securing a sustainable solution can there be a reduction in poverty, as well as an increase in food security.

If this wasn't enough to make you consider whether to continue to eat fish or not, then just look at the health implications. Studies show that pretty much all fish contain mercury. When humans eat the fish, the mercury transfers into our bodies, where it builds up. Studies have shown that those who eat plenty of fish have mercury levels in their bodies which exceed the maximum recommend amount[36]. When we consider that elevated levels of mercury can lead to a number of medical issues, including damage to the nervous system and the brain, especially on unborn foetuses and young children, then this is surely something we should know.

One alternative use of the oceans for food is with the farming of edible sea plants. Edible seaweed, for example, is one of the fastest growing plants in the world, absorbing carbon dioxide and producing nutritious food. It also supports other ocean life, such as shellfish. Around a half of the seaweed's weight is bio-oil too. This can be used rather than fossil fuels for vehicles such as cars and planes. When done properly, this not only produces healthy, nutritious food, it also increases fish stocks and biodiversity, whilst requiring no extra water, no deforestation, no fertiliser and no heavy machinery. It can also be implemented at many different depths of water, increasing the amount of space available.

Food Waste

Over one-quarter of all food around the world is wasted each year[37]! This can happen in a variety of ways all the way along the supply chain. In total, a full six percent of the world's total greenhouse gas emissions come from food waste. This is around three times that from all the world's aircraft. Food waste also leads to far more land and water use than would be needed otherwise. This waste also costs over two and a half trillion dollars per year[38]. Whichever way you look at this, it is something that needs to be reduced.

The majority of waste comes from poor storage and handling techniques, lack of refrigeration, spoilage in transport and processing, food being thrown away due to how it looks, reaching 'best before' dates and from customers. The more that growers, suppliers and consumers can do on this the better.

Whilst food waste needs to be reduced, the remainder can be put to use. Currently most food waste ends up in landfill or burned. It contains huge amounts of microbes and nutrients so why is so much of it going to waste? Instead, all food waste should be collected and composted. This of course includes compost toilets – our own waste should stay in the loop and made use of. Once more, we need to mimic nature. Of course, this will lead to reductions in profits for the huge multinationals that produce so much current fertiliser, but some might say that our future is worth more than their profits…

Drinks

So far, we have just concentrated on food, but it would be wrong to not also consider drinks too. Probably the biggest observation made so far is that animal products have a significantly more negative environmental impact, in every aspect, than plant-based products. To see if this is also true for drinks, let's now compare the main animal-based drink, namely milk, to the alternatives.

The evidence tells us that dairy milk emits over four and half more greenhouse gases than almond milk, over three and half more times that oak milk, over three times more than soymilk and over two and a half times more than rice milk. When we compare land use then the figures are even worse for dairy milk. It requires thirty times more land than rice milk, eighteen times more than almond milk, almost thirteen times more than soymilk, and over eleven times more than oat milk. Lastly, if we look at water use then dairy needs over twenty-two times more than soymilk, thirteen times more than oat milk, over twice as much as rice milk and over one and a half times more than almond milk[39].

In summary, dairy milk is by far the worse, environmentally, of the milks available on the market. One way then, to reduce your negative environmental impact, is to switch to non-dairy milk.

As for water, it is estimated that the environmental impact of bottled water is three-thousand-five-hundred times that of tap water[40]. This includes an impact on ecosystems of one-thousand-four-hundred times greater. And it is so much more expensive. The advertising companies have done a very good job of convincing us to spend our money on harming our planet. Also remember that bottled water companies don't make water, they make plastic.

Food and Disease

In 2020 our awareness of zoonotic diseases changed forever. Zoonotic diseases are those which have passed from other animal species to humans. Three major diseases of our times, namely HIV/AIDS, Ebola and, of course, Covid-19, are all zoonotic. Other examples of zoonotic diseases include avian flu, swine flu, Spanish flu, which devastated the world in 1918, and the Black Death which is still the deadliest pandemic in human history. This is not all however, the World Health Organisation estimates that sixty-one percent of current human diseases are zoonotic, including seventy-five percent of new diseases which have been discovered in the last decade[41].

Many of these have had devastating consequences, but why are they occurring? When human beings are under stress, our immunity drops and the levels of viruses, bacteria, etc, increase. The same happens with other animals. When animals are removed from their natural habitats, such as due to deforestation, they can become stressed, thus increasing the risk that they can spread disease. Humans, often being the reason why they have lost their habitat, then come into contact with these animals, and have the diseases passed onto them.

Whilst is often wild animals that get bad publicity regarding zoonotic diseases, livestock and pets actually carry eight times more viruses that can be shared with humans, than wild mammals[42]. Most transmission from wild animals occurs from those which have adapted to living in urban areas, such as rodents, bats and primates. Human activities such as animal agriculture, deforestation, hunting and animal trade directly increase the chances of zoonotic diseases being passed on to humans.

If we are serious about reducing the chances of devastating zoonotic diseases in the future, then we need to rethink our relationship with wild and domesticated animals. Killing animals could never be the answer, and so we need to look at how we treat animals, both wild and domesticated. Do we need to keep animals in cramped conditions and then slaughter them? Do we need to destroy their habitats, hunt them, and sell them for money? Because if we do, knowing what we know about the causes and potential consequences of zoonotic diseases, then we are asking for trouble. Even for those who currently don't care about environmental issues, we have all seen the economic impacts of Covid-19. Surely anything to reduce the chances of that happening again is worth doing.

The other major issue with disease comes from the use of antibiotics. Antibiotics are drugs which are used to treat bacterial infections. In animal agriculture, antibiotics are used to prevent and treat disease, as well as to enhance animal growth. This is happening in huge numbers; in the USA, for example, 4 times as many antibiotics are used in agriculture than to treat humans[43]. So, what are the consequences?

The overuse of antibiotics in animal agriculture has led, and will continue to lead, to the creation of bacteria that are resistant to antibiotics. These so-called superbugs can then be passed to humans. Antibiotics have been extremely successful in the treatment of bacterial illnesses, but increasing antibiotic resistant diseases, such as MRSA, could have huge consequences for human health across the world. The WHO reports that at least seven hundred thousand people die from superbugs each year[44] and antibiotic resistance is one of the greatest threats to human health.

If that's not enough, then know that the antibiotics used in animal agriculture can be ingested by humans consuming the animal products, and that up to ninety percent of antibiotics pumped into animals is excreted, thus being released in the manure and ending up in water supplies[45].

Some countries have reduced their use of antibiotics within animal agriculture, but this is not happening fast enough. Yet again we are seeing the devastating effects of animal agriculture. In humans, antibiotics are only used to treat disease, and it should be no different with farmed animals.

Of course, the big pharmaceutical companies, which make huge amounts of money from the sale of these antibiotics, are extremely powerful, are very influential lobbyers of government, and have the knowhow to persuade farmers to keep buying their produce. Until 'big pharma', like fossil fuel companies, are removed from politics and influence, not enough will change.

Grow Your Own

Across the world some people are fortunate enough to have some of their own land, whether owned or rented, be that a garden or allotment. Growing your own food does not only give us a sense of doing something positive, but you are choosing what to grow, within suitability of soil and location, so waste is far less likely. Also, you know how much water, fertiliser and other chemicals is used. All this helps us recognise that we ourselves are part of nature and the world is not a separate resource to be exploited. It will also help us see that there is another way, particularly in certain parts of the world; we don't have to rely upon a drive to a large supermarket for everything.

Added to this, if we also collect rain in water butts, not only can we use it for watering plants, thus saving on other water and energy, but it also reduces the amount of water going into the sewers. This is particularly important in areas more prone to flooding.

The Cost of Food

Currently, in many parts of the world, sausages are cheaper than apples. How can this be true if growing apples is more efficient than animal-based products? Here we see a similar situation as with fossil fuels. The livestock industry is built upon or at least greatly supported by large government subsidies[46]. On a level playing field meat would be far more expensive.

As we have seen, animal agriculture has a large negative impact on our environment. Imagine if these companies then had to pay for the clean-up of the environment. Instead, who pays for this? Either the tax-payer or it is not done at all. Is this ok? How do you feel about this?

Again, like with the fossil fuel industry, the meat, fish, and dairy industries are powerful lobbyers. Political parties gain significant support from this industry. If these subsidies were removed and, indeed, a tax was added on animal products, much like those on tobacco and alcohol, then simple economics would mean that people would transition more from animal to plant-based products.

Also, similar to fossil fuels, many financial institutions, such as banks, invest heavily in this industry, including in those vast multinationals that may make money, but cause a significant proportion of the damage done by animal agriculture. Are these the institutions that you want to support?

Effect of Climate Change on Food

As a species, food security, along with air and water, is the most important issue for our survival. Let us consider how well this is working now, and what may change in the future.

The current food system across the planet produces significantly more food than we need. We have seen that this, in part, is due to the high proportion of food which is wasted every day. But since so much food is produced despite the waste, there should be more than enough for everyone, but is it distributed fairly?

Well, over eight-hundred-million people across the world are undernourished[47], whereas around two-billion adults are overweight or obese[48]. How can this be right? Clearly the system is not working for everyone. Therefore, it is important that we look at potential changes to make things better. As part of this we must also look at the impact of the changes that we have made on the environment.

Climate Change is already leading to increases in temperatures, greater variation in rainfall and a higher frequency in droughts, floods, and hurricane-like events, as well as in increase in pests. All of these are affecting food security in many locations. Madagascar, for example, is currently facing the world's first famine, brought upon purely by Climate Change[49]. This has caused food insecurity for over a million people, with around four hundred thousand facing famine. Indeed, the situation is so grave that locals have had to resort to eating raw cactus, wild leaves, and locusts, just to survive. Just imagine if that was you and your loved ones.

In some areas, plants will respond positively to increasing temperatures and higher levels of carbon dioxide. This means that they will grow faster, more efficiently, and with longer growing seasons. Other areas, however, will just not be able to grow the plants that have been there for generations. Crops, and grasses, will no longer survive. The conditions may even become so bad that nothing can be grown at all. This leads to a reduction in food security. It will also cause an increase in prices, putting the less materially wealthy at even further risk.

The keys to this are biodiversity and adaptation. Biodiversity in food brings resistance to the effects of disease and insect infestations, whereas adaptations must take place to deal with the changing temperature, levels of carbon dioxide, soil, and precipitation.

At this stage, some may say that since there will be issues with growing certain plants then we should look to breed even more animals for food. This, however, ignores the fact that these animals need to be fed too. If we bred a greater number of farmed animals then the total amount of plants required would increase, leading to a reduction in our own food security.

Within this we also have the responsibility to ensure that all of us have adequate and suitable nutrition. This means supporting those receiving not only too little, but also too much. We know that both can lead to serious health consequences, and it is important to ensure that the system developed is a system for all.

Summary

The message is clear; in our modern world, food choices are having a significant negative impact. This is seen in terms of carbon emissions, pollution, land use, water, biodiversity and even food security, but it doesn't have to be like this. Since we understand that our food choices make a such a difference, then we must lessen how much meat, fish, and dairy we are consuming. This isn't saying that everyone needs to go vegan overnight, but that each time you choose a plant-based product over an animal-based one then a positive choice is made.

If we think about the science of this, then it makes sense. In order to produce animal products, we must first grow plants which the animals then consume. These animals obviously use a lot of the energy from this food during their lives. This means that a lot of the original energy from the plants is unavailable to the humans who eventually consume the animal. Animal-based products are then far less efficient than plant-based ones. Yes, there are many animals which eat meat and fish in nature, and they play a vital role within each ecosystem, but anyone would be hard pushed to argue that our current agricultural system is anything like what happens in the natural world. It's not the same.

Of course, meat can taste good, but do our taste buds justify the extra environmental damage? There are, of course, other reasons for reducing your use of animal-based products too, such as animal welfare and health. For example, humane killing means killing with compassion. How do you kill an animal that wants to live with compassion?

The food we eat has been ingrained into us since birth. It is part of our culture and our heritage. But this doesn't mean that as times change then we shouldn't change with them. Many of us have been bombarded by messages that it is manly to eat meat, that we need meat for protein and to grow muscles, but a growing number of elite athletes and bodybuilders are crushing this myth. There is also growing medical evidence directly linking animal-products to the likes of heart disease, diabetes and certain types of cancer[50].

There may well be reasons why some animal products should still be used but before doing so, look at the evidence. We should also never judge those who continue to both produce and consume animal-based products. None of us have walked the life of another and so the aim should always be to work with, support and educate each other, rather than judge and condemn. This very much includes those who currently rely upon their animals to survive.

It is also important to recognise that people live in a great variety of situations across the world. This chapter is not saying that those people who have no alternatives but to eat animal products shouldn't continue to do so. Indigenous populations who live a sustainable lifestyle should absolutely carry on what they are doing. They are not the ones causing the problem. The issues discussed here are caused by the current systems in place across many parts of the world which make animal products unnaturally cheap relative to those which are plant-based and rely upon the continued exploitation of our natural resources. As we have seen, many modern agricultural practices are simply not sustainable. We all need to be part of that change.

We will look at what we can do both as part of a group, and as individuals, in Chapter Nine. For now, switching to a plant-based diet is probably the biggest positive difference environmentally that an individual can make, so ask yourself, what would someone who wants to mitigate Climate Change eat? Remember, this is our climate, our future, and there is a better way.

CHAPTER 5

WEAPONS OF MASS CONSUMPTION

Since the moment we are born, we are continually bombarded with messages, some far more subtle and sophisticated than others, on how much better our lives will be, if only we had this or that product. We have literally become weapons of mass consumption in a bid to make our lives better, to make ourselves feel better, to fill a void which exists in our lives, as we succumb to the powers of advertising.

In 1967, the space race between the USA and the then USSR was in full flow. One consequence of this was a photo that forever changed our understanding of planet Earth, namely the first colour, full-disk, photo of Earth[1]. For many this was a game changer. It showed our planet to be the finite, limited. No longer to be considered an infinite resource to be plundered. But this has not stopped our consuming ways.

Our levels of consumption are now at such an extent that Earth Overshoot Day, the day in the year in which we reach the point that we have consumed more resources that year than the Earth can regenerate, moves ever earlier. In 2021 this day fell on July 29th [2]! That is, in just seven months we consumed as much as the Earth can sustain in a year. Clearly this is a situation that cannot continue. Added to this, we are living within political and economic systems which push for ever more growth year on year, with an intended and deliberate increase in our plundering of the Earth's resources. It just doesn't make sense. We will look more at the systems in Chapter Eight, but for now let us look at some of our impacts.

When we think about the environmental impact that we are having on our planet then it is normal to consider human created products, such as the plastic straw which infamously needed to be removed from the nose of a sea turtle[3]. Clearly that product should never have been there, and such videos of such events raise significant awareness. This is just part of the manufacture, use and disposal of a variety of material items which are negatively impacting upon our environment. But so many of us like our material items, so what can be done?

In this chapter we will look at the impact of a variety of different material products and what can be done to reduce the negative effects that they are having. This doesn't just mean upon wildlife but also directly on ourselves. Whilst, as always, everything is interconnected, it seems sensible here to look at each material product individually.

Plastics

Plastics often get a bad press environmentally, and generally rightly so. Therefore, it is important that we understand what they are, why they are used, their impact, and what we can do about it.

A plastic is any synthetic or semisynthetic organic polymer. What? Well, this basically just means that all plastics contain long chains of carbon atoms (a few have silicon instead). Different plastics have different length chains of carbon, as well as different atoms connected to these chains, with hydrogen being the most common. Another way to look at plastics is that they are a type of material that can be easily put into any shape and then maintain that shape.

The first plastic was created in 1869[4], meaning that until 150 years ago, there was no plastic anywhere on Earth. Even then, it took many years for the use of plastics to really take off. A variety of different plastics were produced over time, but it was really during the second world war when the number of plastics really started to increase. Since that time, plastic production has increased from very little to over three-hundred-and eighty million metric tonnes per year[5]!

Plastics were originally seen as a wonder product; they were cheap, easily moulded into any shape, could be soft or hard, and last a long time. Different plastics could then replace a whole host of other materials. Plastic bags, for example, were cheaper, lighter, and stronger than their paper counterparts. Not only that but many saw them as a real plus environmentally. It takes more energy, and therefore more carbon emissions, to produce a paper bag than it does one made from plastic. Win, win! Or was it? Whilst an explosion of plastics occurred, they had a darker side. Possibly the biggest issue regarding plastics is that they last so darn long. Most of these items, such as plastic bottles, will remain within our environment for hundreds of years before they break down[6]. The reason for this is because these are not natural materials. This means that no bacteria, or other organism, has evolved to break down these plastics. The only breakdown that does occur is mostly due to light, friction and by being eaten by animals.

Even though it clearly takes far too long for plastics to break down, they will do so eventually, right? So maybe it's not too bad. Let's look at what they break down into. As the pieces of plastics become smaller and smaller, they first become the aptly named microplastics. These are still plastics, but no more than about five millimetres in length. Then what happens with these? Well, not a lot, not really. They might become a bit smaller again and become 'nanoplastics' but it is still plastic. Light can cause some of them to decay further, but they then release gases such methane[7], which as we saw previously, is a very significant greenhouse gas.

It should also be recognised that, in many ways, microplastics and nanoplastics are worse that larger plastics. Being so small and light, they move easily from place to place. They have even been found in even the most remote places on Earth, such as within

Antarctic ice[8]. Being so small it is also very easy for them to be taken up into the food chain.

Plastic debris was first found within our oceans in the early 1960s. As time went on this has increased significantly, with around eight million metric tonnes now going into the oceans every year[9]. That's like dumping a lorry load of plastic into the oceans every minute! So, let's look at how it gets there.

One significant contributor is fishing gear, such as nets, lines, and traps. This so-called 'ghost gear' contributes up to one million tonnes of ocean plastic every year[10]. Whilst there are ways to reduce this, this is clearly a consequence of fishing. Therefore, by supporting this industry we are, inadvertently, but in a very real way, adding to the plastic in the oceans.

Plastic also enters our oceans after being blown off landfill sites. It then makes its way into our drains and rivers, and then the oceans. This is typically plastic bottles, other containers, and straws, but also microbeads from cosmetics, clothing fibres and many others. This is then taken by winds and ocean currents all around the world, from deserted islands to the bottom of the Mariana Trench, the deepest part of the sea[11].

There are even regions of the oceans, such as the Great Pacific Garbage Patch. The ocean region contains over a trillion pieces of plastic[12]. Whilst it is more like a cloudy soup rather than a visible mass of floating debris, it is clearly very disturbing. Also note that this is just one of many so-called ocean garbage patches around the world.

Some of the worst plastic pollution comes in the form of nurdles. These are small pellets which are used to make most plastic products. Huge numbers of them are shipped across the world. Accidents then occur, and hundreds of thousands of tonnes of them are spilt into our waters every single year![13]

Not only have these washed up on our coastlines, sometimes up to two metres high, but they can easily be mistaken for food. They have been found in the stomachs and mouths of dolphins, fish, seabirds, and other wildlife. What's more, they are also effective in transporting bacteria, toxins, and other pollutants, from one part of the ocean to another. These substances then go into whichever species happens to eat them. In certain parts of the world this is also having a detrimental effect on the livelihoods of locals, including those who benefit from fishing and tourism. Not only is all of this plastic in our oceans extremely ugly, it has a significant impact upon marine life. For example, one whale was found to have eight-eight pounds, that's over six stone, or forty kilograms, of plastic filling its stomach[14]. This caused the whale to vomit blood and subsequently die a most painful death. How can this be ok?

The evidence is no better when we come to other species. Around one million sea birds and one hundred thousand marine mammals die from plastic pollution every year[15]. It is also found that pretty much every baby sea turtle in the oceans has plastic in its stomach. This is all a direct consequence of our production, use and disposal of items which we often only use for a few minutes. Is this something we can continue to justify?

This is clearly horrendous, but the same is happening to you and I on a daily basis. The atmosphere takes up large numbers of microplastics, causing the very rain that we rely upon so much, to contain plastic within it. Indeed, it has been determined that over one thousand metric tonnes of microplastics, equivalent to over one hundred and twenty million plastic water bottles, fall as rain each year, onto just six percent of the USA[16]. If the same is happening elsewhere around the globe, which is likely to be true to a lesser or greater extent, then these numbers become even more horrifying.

Well, what does this mean? Firstly, the very water that we all drink contains plastics. And it's no good thinking you're just going to drink bottled water instead, as on average, this contains around fifty percent more[17]! It goes without saying that all other drinks, being mainly water, also contain these plastics too. There is no getting away from it.

Microplastics make their way into our bodies via other means too. In both the oceans and on land, microplastics are ingested by many small organisms such as bacteria, algae, and plankton. These are then eaten by larger creatures, such as fish, until it ends up on our plates. This means that those of us who eat fish and shellfish take the greatest amounts of plastic into our bodies. The rest of us, however, are not safe either. Microplastics from wind and rain fall onto agricultural land. So, whilst more studies need to be done, it seems that there is a reasonable chance that all our food and drink now contain some of these plastics.

To understand better which foods are likely to have a greater proportion of microplastics then we should understand two phenomena known as bio-accumulation and biomagnification.

Bioaccumulation is where small amounts of a pollutant is typically taken in by plants. Since this pollutant isn't broken down easily, it remains within the plant. The plant then takes in more of this pollutant from its surroundings, causing the amount of that pollutant within that plant to increase, i.e., to accumulate.

Then we have biomagnification. When these plants are eaten, these pollutants are passed onto the next species. Again, the pollutant doesn't break down within their bodies. These animals eat a significant amount from many different plants, leading to them acquiring a higher proportion of the pollutant within their bodies than the plant did. In other words, the amount of pollutant has been magnified from one species to the next. If these plant eaters are then eaten by predators, then the process continues further. This causes those at the top of the food chain to have the highest proportion of pollutant in their bodies. Guess where we fit.

This also means that, on average, plant-based food will also have a lower proportion microplastics than animal products. Something else to consider when you are choosing what to eat.

Studies suggest that the average person ingests the amount of plastic in one credit card, per week[18]! Fancy eating a credit card each week? No, me neither! But does all this really matter? Is it actually harmful to us? Well, you may be surprised to hear that the research on this is currently fairly minimal, but they quite possibly cause inflammation and even increased chances of certain cancers[19]. When the pieces are small enough, they can even pass through into the blood stream, then onto the rest of the body. Whilst the data on humans is still unclear, one study looking at nanoplastics in fish brains noted a measurable change in their behaviour[20]. Do you know what, it's a no from me!

If you weren't already, then you are now, quite possibly, thinking that our use of plastics is a really bad idea. So, we should look at why we are still using them. Remember that in many ways, plastics are amazing. As mentioned above, they are cheap, strong, easy to use and mould, and last a very long time. If used correctly they could potentially be great, but this is clearly not what is happening. So, what is really going on?

To understand this, let's have a look at where they all come from. The vast majority of plastics are made from fossil fuels. This generates massive revenue for the fossil fuel industry. We have looked at some the issues regarding such corporations in Chapter Three, so I won't repeat myself here, but it is safe to say that they are an extremely powerful lobbying power around the world. Is this a factor?

When it comes to plastics, like with other items and services, the more we use, the greater the profits. We have seen that the fossil fuel industry receives vast government subsidies. This then makes the cost of these plastics artificially low. Indeed, they are so cheap that consumers will happily accept the use of single use plastics, rather than their more expensive alternatives. Clearly, as noted in Chapter Three, these subsidies need to stop if we are going to have real change.

This also comes with a huge amount of propaganda. These companies are very quick to blame individual behaviour rather than their own actions. They make a big point about how it is not the plastics themselves that are the problem, but how we use and dispose of them. Some have made minor improvements in recycling and waste collection, but they know that this won't make a significant difference. It does, however, allow them to portray themselves as the good guys, whilst, in reality, being nothing of the sort.

This propaganda is so strong that many people around the world consider the use of plastics as a non-issue given that we can recycle them. In practice, however, many plastics are very difficult to recycle. Also, with those which can be recycled, not only is energy required to do this, but for the vast majority of plastics this never happens[21].

Indeed, a massive seventy-nine percent of all used plastics are now either littering the environment or in landfill. Of the remaining, twelve percent have been incinerated, with just nine percent recycled. It's even worse when we consider the fact that plastic degrades within the recycling process. This means that it can only be used for less useful items[22]. It should be noted that technology is advancing in the field of plastic recycling; it is now possible to recycle any plastic[23], but this is not currently available on a large scale.

All this tells us that relying on recycling is not the answer. Instead, apart from with some medical use, single use plastics should be banned. Otherwise, only those which can be continually reutilised have a justifiable place in our world.

Let's now look at who carries out the plastic recycling which does currently occur. Surely each country does their own. The wealthier countries are surely responsible enough to deal with their own waste, aren't they? Well, sadly not. Until 2018 huge amounts of plastics were shipped to China. Then the 'National Sword' import ban began, meaning China would no longer accept these items[24]. You would hope that this would have been a wake-up call for wealthy 'Western' countries and they would instead deal with the waste themselves. But no.

Instead, this 'plastic' waste is exported to other countries, such as Vietnam, Malaysia and Thailand[25]. These places don't always have the infrastructure to deal with all the waste and much 'recycling' ends up in rivers or is set on fire. In this process, toxic fumes are emitted. Those who live near these incinerators have been observed to experience skin rashes, lung damage and breathing difficulties. This had led to children being stopped from playing outside their homes with these areas. Those who make money from the heat and electricity generated from this burning also discourage the likes of recycling since they want the waste to make money. Workers in these plants are also often so poor that they have to accept working in such unhealthy conditions, just to survive.

Some individuals in the 'Western' world have even stated that there is no point in doing anything about this waste since there is so much single use plastic in other parts of the world. Not only is this incredibly unhelpful, it also doesn't look at the core issue. The truth is that just twenty corporations produce well over half of all plastics, with many of these being from the 'West'[26]. Many of the products they produce are also combined with other materials in so-called multilayer packaging. This makes any recycling extremely difficult.

A prevailing attitude still seeks to blame others, however. For example, some point the finger at Asian countries, citing poor waste management. But what is the overlying issue? Well, not only are waste plastics transported there from around the world, but aggressive marketing from hugely powerful corporations has turned people away from the containers they would have used over generations. Instead, people have been led to rely upon single use plastics. All to make profits, and with the full backing of governments.

What we should see from this is that a move to zero waste is the only sensible way forward. Recycling may have a place, but it is nowhere near as effective as other Rs, namely, in order; Rethink, Refuse, Reduce, Reuse, Repurpose, Repair, Refurbish, Regift and Rot. We must regain our respect and value for material items.

To help combat this, some countries have implemented costs for any single use plastic. When this has happened, their use has reduced. In the UK for example, since the introduction of a five pence cost for a plastic bag in 2015, the number of such bags reduced from about seven and a half billion to under two billion per year by 2018. Whilst this is clearly still too many, it does show that policies can work.

In other countries there has been a complete ban on such products. In Kenya, by 2017, plastic pollution was rife, clogging up drains and piling up in neighbourhoods. Single use plastic bags were then banned altogether, and by 2020, previously filthy areas were clean once more[27].

Hopefully we can now see that, whilst individual action is important, a far greater change occurs when the large multinational corporations and the governments are held to account. Imagine a world where those who created the plastics in the first place were those who were also responsible for their proper disposal. If these corporations were liable for the whole life of each plastic item, alongside the removal of their subsidies, the difference we would see would be extraordinary.

Sadly, many governments will only go for real change when their voting base turns against them. Also, the corporations will only change when people refuse to purchase their items. Significant raising of awareness and lobbying is required if we are going to get where we need to be.

This isn't saying that all plastic production should stop,

since it can be such a useful product. Rather, the systems need to change so that any such plastics are either used in medicine, or are genuinely used for life, or indeed, for generations.

Now there are alternatives, such as plastics which don't come from fossil fuels. These are plant-based plastics such as PLA (Polylactic acid)[28]. One advantage of PLAs is that they genuinely break down. They are often marketed as being compostable within twelve weeks. Sadly, this is not really true. It is possible to break them down in this time frame inside an industrial composter, as long at the correct microbes and temperatures are used. Most PLAs, however, end up in landfill. In these they decay over years or even decades. So not the advertised twelve weeks.

They do genuinely decay into the water and carbon dioxide that the plant took in as it grew, and so are much better environmentally than traditional fossil fuel plastics, but in further encouraging the use of single use plastics, they cannot be the answer. It should also be noted that they are often mistaken for other plastics and so are put into, and contaminate, other recycling. Whilst PLAs aren't currently recycled themselves, scientists have now developed a method to do so which is so effective that they can be recycled hundreds of times[29]. This, however, is currently not in practice. Some plastic bottle companies are switching to aluminium instead. In many ways this is better, but it takes more energy to create a can than it does a plastic bottle, and thus greater emissions. This is, therefore, not a long-term solution. Indeed, any of the alternative single use solutions should be seen as a form of greenwashing, and only serve to support corporations in making profit.

When we look at the big picture, we can see that zero-waste management is not only better environmentally but also economically. No longer would there be vast amounts of waste to collect, sort, process, put in landfill, burn, and recycle. No longer would huge subsidies need to be paid to the fossil fuel companies. It will need a shift in behaviour. Customers will need to bring their own containers, as well as return reusable packaging to stores, but surely this is minor compared to the alternative.

Clothes

Let's now consider the environmental impact of the items that we wear. Apart from fossil fuels and animal agriculture, the clothing industry is deemed to be one of the most environmentally damaging on the planet[30], but why? To understand this, we must look at the energy and chemicals required, and the final disposal of the items.

The most used natural fibre in clothing is cotton. Nice, natural, healthy, plant-based product, so what's the issue? Well, whilst just around two-and-a-half percent of the world's current cropland is used to grow cotton, it consumes a massive seven percent of all herbicides, and a staggering sixteen percent of all insecticides[31]! For this reason, cotton is often referred to as the world's dirtiest crop. There is also the more obvious use of dyes. These chemicals not only pollute the air and waterways, the dust and tiny fibres they release can severely damage the lungs of the workforce.

If this wasn't bad enough, a huge amount of water is also required. For example, the manufacture of a single t-shirt requires the use of up to two-thousand-seven-hundred litres of water[32]! So much fresh water is required that a number of lakes across the world have completely dried up. One example is the Aral Sea in Uzbekistan[33]. Once the fourth largest lake in the world, it is now virtually gone. Just imagine if you lived near to it and relied upon its water for all the uses that so many in other parts of the world take for granted.

When we look at the energy use, the making of a single pair of jeans can lead to the emission of over seventy pounds of carbon dioxide[34]. Indeed, it is estimated that a massive ten percent of all carbon emissions are due to the clothing industry[35]. This isn't helped by the one-hundred-and-twenty-million trees which are required to be cut down each year[36].

Many clothing items are also made from artificial materials such as polyester. This is a plastic made from fossil fuels. It takes around seventy million barrels of oil to create all the polyester used each year[37]. This causes the same issues regarding plastics that we discussed above.

The clothes also need to be washed and dried at regular intervals. Every time this happens, tiny fibres break off and enter the ecosystem. In many cases, electricity is also used to wash and even dry the clothes, leading to further emissions.

On top of this, many fast fashion companies pursue a strategy known as planned obsolescence. This means that they design garments such that they become unfashionable, wear out, lose shape or fall apart easily. This is done to ensure consumers have a regular need for more clothes. All in the name of profit.

Even if you don't care about environmental issues (which you probably do if you are reading this book) then how about the fact that many large clothing companies have been reported to still use sweatshops and even child labour[38].

And as much as we want to just blame the corporations, or those running the factories in far-away lands, we must recognise the role of consumers, with their insatiable desire for new, cheap clothing. If you try making your own clothes, for example, you'll realise just how ridiculously low some clothing prices are. Ok, so they are often mass produced, and the fabrics may be cheaper elsewhere, but when you factor in all the additional costs, as well as the profits the companies make, how can the clothes possibly be so cheap? Just how bad are the materials? How much, or rather how little, are the workers being paid? How are they being treated?

It is also worth considering that on each item of clothing, every seam, every detail, was made by a fellow human being. Even with a modern machine, someone still uses their skill to work it. If we think about the person who made it, respect their work, imagine them and their story, then our treatment of such items will change. This thinking extends to all.

And then there is the treatment of unwanted items. Many brands deliberately overproduce, to ensure that they never run out. This leads to places such as parts of Africa[39], and the Atacama Desert in Chile[40] being used as huge dumping grounds. Just imagine if this happened in your neighbourhood. But we need to wear clothes, so what's the answer?

Well first of all, when thinking about purchasing new clothes, consider both the environmental impact and whether you really need them. Then, if you do, why not check what you already have. In the UK, for example, the average person virtually never wears an estimated thirty percent (!) of the clothes that they already own[41]. You could just wear these clothes, or at least donate or sell them second hand.

If you need to buy new clothes then why not look at the following options; charity shops, second-hand clothing stores, hire the items, attend clothes swap events, or research which clothing companies are genuinely (as opposed to just saying they are) more environmentally friendly than others. Washing clothes only when they genuinely need it and hanging your clothes out to dry rather than using driers will also help. You could also make a commitment to only buy a certain number of items per year, and no brand-new items.

Building Materials

Two materials which come to mind when we think of building materials, are cement and concrete. Cement is a powder made from a mix of the likes of clay limestone, sand, and shale. When combined with water it becomes a binding agent and is often used to combine sand and gravel, making concrete. The concrete is then used to make the likes of roads, bridges, and buildings.

After water, cement is the second most used material in the world, with over four billion tonnes produced every year[42]. Clearly it is extremely useful, but it comes at a cost. For every kilogram of cement produced, a kilogram of carbon dioxide is released. This causes around eight percent of all carbon emissions[43].

Firstly, energy is required to extract and transport the raw materials. The limestone (calcium carbonate) is then heated and breaks down into the calcium oxide, whilst releasing carbon dioxide. The energy used can be relatively low carbon, but there is currently no way to avoid the requirement of limestone, and thus the release of huge amounts of carbon dioxide. Options are being trialled to get around this but there is currently nothing suitable on a large scale.

Another material used extensively in construction is steel, with just under two billion tonnes produced worldwide each year[44]. Steel is strong, durable, relatively economical, and can be made in a short amount of time. It is also widely recycled. For this reason, it is sometimes marketed as environmentally friendly. Sadly, this does not paint the whole picture.

The mining, transport, and production of this steel produces about two kilograms of carbon dioxide, for every kilogram of steel[45], causing around seven percent of all carbon emissions[46]. To produce the steel, a high purity type of coal, called coking coal, is also used. This is heated without oxygen to remove impurities, producing an almost pure lump of carbon called coke. This coke is then heated with iron ore, and other materials, creating steel. The importance of the coke is that it removes oxygen from the iron ore, and combines it with the coke's carbon, making vast amounts of carbon dioxide.

But is there another way? Can steel be 'decarbonised'? Well firstly, about twenty-five percent of global steel production comes from recycled steel each year[47]. Some steel is also reused. But what about the rest? Well research shows that hydrogen, rather than coke, can now be used to produce steel. In Sweden, for example, green hydrogen has been utilised to make steel without the direct creation of carbon dioxide[48].

There are issues, however. Modern furnaces are designed for use with carbon, not hydrogen, so would have to be adapted. Also, the use of hydrogen is currently notably more expensive than using coke. I haven't found anything on subsidies here, but it seems reasonable to consider what would happen if subsidies went to green hydrogen rather than to the fossil fuel industry.

We could also, to some extent at least, use wood instead of the aforementioned materials. This way, rather than causing huge emissions, the houses would be stores of carbon. Obviously, the large scale cutting down of trees is never the desired solution, but, as with all things, we must weigh up the various pros and cons.

Another way forward is to build and adjust existing buildings so as to make them as low carbon as possible. This may include the following; on site solar energy, daylight rather that electric lights, designs which encourage people to walk, high levels of insulation, electrochromic glass/smart glass, the collection and use rainwater, composting organic waste such as sewage, incorporating green roofs, and having efficient heating and cooling systems.

Overall, though, the best way right now to reduce emissions from the likes of cement and steel is to use less. But we need new houses and roads, right? Well, there may well be a need, but is this really as much as is suggested. For example, in England alone, at the end of 2019, there were almost six-hundred-and-fifty-thousand empty homes. Two-hundred-and-twenty-five-thousand of these had been empty for at least six months. Another seven-hundred-and-seventy-thousand households reported owning a second home, with eight percent of these having more than one extra. Whilst some of these are rented out, a great many are not. This means that there are huge numbers of houses across England which are simply empty for most of the year[49].

Now, whether or not you agree with the idea of second homes is one thing but consider this; around one-hundred-and-thirty-thousand houses are built in England each year[50]. Even this is not deemed to be enough. Blaming those who have second homes helps no one, but imagine a system in which you couldn't do so. Instead, those homes could be lived in by those in need. Consider not only the positive impact on their lives, but also the reduction that this would have on the need for new homes, and therefore the slash in associated carbon emissions.

There is also some debate on whether we need new roads. Of course, if fewer new homes were built then fewer new roads would be needed to access them. Also, roads are often built to deal with excess traffic. If instead there was a far greater emphasis on public transport, then there would be fewer cars on the road anyway, and so there wouldn't be the need for extra roads this way either.

Water

One material that we all need is water. In Chapter Four we looked its use, so now let's focus on water treatment and the energy involved.

When we think of water, it is often considered to be a (relatively) free resource. Whilst we have all heard that we shouldn't waste this water, unless in a time of drought, it's really not a priority for many. Indeed, some of us never consider the energy involved.

To move water from once place to another, it needs to be pumped along pipes. Water is a surprisingly dense material, meaning that even a relatively small volume is heavy; try exercising with a one-gallon container of water! What this means is that a lot of energy is required to distribute the water. The greater this energy use, the greater the subsequent carbon emissions.

On top of this, it is estimated that over three billion litres of this water are lost to leaks in England and Wales each year, which is around twenty percent of the total supply[51]. This means that huge amounts of energy are used to distribute water that never arrives at the intended destination. When we consider that this is likely to be happening all over the world, then the total losses are enormous.

We also need to consider the water treatment process before it comes into our homes. After being collected and stored, the water is treated before use. This includes processes such as filtration and the addition of chlorine. All of this requires more energy, materials, and chemicals.

And what about the wastewater which ends up in our rivers and seas? We discussed issues relating to chemicals from agriculture in Chapter Four, but other industries also contribute to toxic chemicals entering our waterways. In some places around the world, industrial waste is just dumped directly into the rivers. Not only that, but in 2020, in England alone, the water companies themselves dumped raw sewage into our rivers around four hundred thousand times! Indeed, this came to a head in the UK in 2021. The government voted against laws which would have insisted that the water companies demonstrate improvements regarding sewage. It makes you wonder whether this has more to do with profits rather than people, with the bosses of England's nine water companies being paid over sixty-five million pounds over five years[52].

All this can have a devastating impact on local wildlife,

including the addition of far too many nutrients, leading to eutrophication. The level of oxygen in the water is also impacted, as well as chemicals directly damaging the growth and reproduction of many species.

Just ask the people of Flint, Michigan, in the USA. This began in 2014, and the drinking water was only declared safe again in 2019. The drinking water had been contaminated with lead and the bacteria which causes Legionnaires' disease. Nine thousand children were supplied with contaminated water for eighteen months! Lead poisoning has been shown to affect the intellectual ability of children, and the bacteria killed twelve people[53]. This is clearly something that cannot be allowed to happen again.

Maybe bottled water is the way then? After all, it is now the most popular bottled drink in the world, with almost four hundred billion litres being purchased every year[54]. That's about fifty litres per person, in one year! Since most people never drink bottled water, how much are the others drinking?

Bottled water is now a multi-billion-pound industry. And they don't even make water – they make plastic bottles. A lot of them. Hundreds and hundreds of billions each year! We have already said enough about the travesty of plastic across the world, so for now let's look at other aspects.

Firstly, why are sales so strong? Well, it's been referred to as 'The Marketing Trick of the Century' for a reason. Basically, its sales rely upon convincing consumers that it is a healthy alternative to sugary drinks. Of course, this is true, but we could drink tap water instead. Let's then just compare these two.

Obviously, there is a significant cost difference, with bottled water being about five-hundred times more expensive. We've also already seen that bottled water contains about fifty percent more plastic than tap water. As for minerals, there is as much variation between different bottled waters as there is between bottled water and tap water. In other words, on average there is no real benefit from drinking bottled water. The exception would be when water, whether tap or bottled, has become contaminated.

Advertisements also suggest that it is natural spring water. Whilst this is true for about half of all bottled water, the rest, shockingly, comes from tap water[55]!

Overall, the environmental impact of bottled water is deemed to be around three-and-a-half-thousand times that of tap water[56]. It is so easy these days to carry around a reusable water bottle and have it refilled. Surely the priority is the environment rather than the pockets of large multinational companies.

There are other ways to save water too. We saw in Chapter Four that certain foods, particularly plant-based products, use far less water than animal products. We can also use the likes of a water butt to capture rainwater to use in toilets and for watering plants. Other ways to reduce water use in the home include low flush toilets, water efficient washing machines, low flow faucets on taps and shower heads, and using a timer to limit shower time.

Waste

There is no doubt that modern life produces huge amounts of waste, indeed over two billion tonnes every single year[57]! There was a time when material resources weren't so cheap, meaning that they were respected, and more effort was made to reuse and repurpose items. Nowadays however, with packaging and so many products themselves relatively cheap, people are buying more, leading to extra waste.

By definition, waste is anything that has to be thrown away, but what does this really mean? In reality of course, there is no 'away', not when it comes to the whole planet. Waste may well be moved from one place to another, buried, sent to another country, or even burned, but it, or the products created from its burning, still exist, they are still there. Oh, it may well be that we can no longer see them, so it may seem like they are no longer our problem, but as we gain a better understanding of the finite nature of our planet, the more we understand that 'away' does not exist.

On top of this, we currently have a mostly linear waste system, which involves removing resources from their natural environment, making a desired product, distributing it, making use of it, then dumping it as waste. Over time this creates more and more waste, including pollutions. In modern life this all happens on a vast scale, creating significant environmental issues, many of which have been discussed previously.

Surely there is a better way. Imagine a world with no discarded waste, no need for bins, no landfill, no need for constantly producing new materials to replace the old. Let's turn to nature for inspiration on how this might work.

Within the natural world there are so many amazing examples of how to deal with waste, all of which are circular. Well known examples include the water, carbon, and nitrogen cycles. In these there is no start and end, but an infinite cycle, leading to no waste. Let's then use this to consider how a circular waste management system could work for us.

In such a system, products are designed so that there is no need or desire to discard them after use. Instead, they are reused, repurposed, repaired, refurbished, regifted, recycled, or composted. This can be as simple as having drinks containers that can be returned for reuse.

But it is more than that. If we are to make real, positive change then an adjustment in our mindset is also required. The natural world is a thing of beauty. Imagine if we recognised this and worked with, rather than against it. Consider the possibilities if we viewed ourselves as part of nature. We should then see that it is no longer a resource to be exploited, but one within which we can live in harmony.

The ten Rs mentioned previously, namely; Rethink, Refuse, Reduce, Reuse, Repurpose, Repair, Refurbish, Regift, Recycle and Rot, are a useful guide, but within all of this we should consider our actions and whether we want to make a positive or a negative impact on the world.

Let's then look at each of these ten Rs to see how they can help:

Rethink; The most important rethinking that needs to take place is from governments and large corporations. This includes moving to a circular waste system. Legislation could ensure that corporations carry out Extended Producer Responsibility (EPR). In other words, those who produce and profit from the products are then responsible for their recovery and reuse. This would encourage these corporations to consider the whole lifetime of an item. This would greatly affect their design, making them longer lasting and easier to fix.

Of course, individual rethinking also plays a part, and we should all consider our own impact. This can be tricky since we are bombarded from birth with the importance of having, doing, and spending more. As we understand this better, we can see that a significant action is applying pressure to our government and any corporation that we deal with. This may be in the form of protest, lobbying, voting, and careful purchasing (or not). This is all supported by being a positive role model.

Refuse; This one is clear in principle; simply refuse to purchase and use certain resources. This may start with the likes of single use plastic bottles, or red meat, and then, over time, permeate through all of our choices.

It sounds like a negative but can lead to creativity, a feel-good factor, and a closer bond with nature.

Reduce; Now, we all have a wish to live, thrive and survive, so we can't refuse everything. What we want is a better world for everyone, just one that doesn't cause irreparable damage. By reducing the number of items we consume, we can consequentially reduce the waste we generate.

Reuse; Wherever possible we can use reusable items only, and then look after and reuse them, over and over again. This will clearly have a marked impact upon the number of purchases we need to make.

Repurpose; Sometimes we no longer have a use for an item. Instead of it just throwing it away, why not get creative. There are so many ideas out there for how to repurpose different items or come up with your own.

Repair; It can be so easy in our modern world, to throw items away and buy a new one, rather than get them repaired. There are a number of reasons why we may do this, including the ridiculous consequence of our current system which is that with things that we cannot repair ourselves, it is often far more expensive to get it repaired compared to buying a new one.

Refurbish; To refurbish an item is to make it as it was when first purchased. This includes repairing but also making it look new again. This is particularly done with electronic items, e.g., mobile phones. Buying refurbished goods reduces energy use and need for new resources. This in turn reduces waste, including the release of toxins into the environment.

Regift; To regift is to give someone something that had previously been gifted to you. We know that huge amounts of money are spent every year on gifts for people who don't want them. Obviously, the ideal is for these items to not be purchased it the first place. Conversations about what you explicitly would like can be tricky but will significantly reduce the chances of getting things you don't want. If you do happen to still get something you don't want, then consider who would appreciate it and give it to them.

Recycle; This is often lauded as making a real difference environmentally, so let's look at this in more detail. One obvious issue is that this process requires energy. Whilst this is noticeably less than that needed to make new materials, we noted in Chapter Three that our use of all energy resources has an environmental impact.

Another significant factor is that, as mentioned with plastics above, some materials degrade during the recycling process. Others are downcycled within the process, meaning that the products made from the recycling are less useful than the original ones. This all means that many products can only be recycled a very limited number of times. Most plastics, for example, can only be recycled no more than twice[58]. Paper can be recycled up to seven times, whilst materials such as steel, aluminium and glass should be able to be recycled as many times as required.

Another issue with recycling is that many people continue to purchase single use items, thinking that it is fine to do so since they will just put them in the recycling. In this way recycling actually supports the production and use of single use items – the very thing that so many people involved in recycling wish to avoid.

We also saw previously that so much so-called plastic recycling is sent to less affluent countries around the world where it is often not dealt with properly and has led to significant environmental and human health issues. It's like some people in wealthier parts of the world use an 'out of sight, out of mind' mantra and don't really care what happens, as long as they don't need to know about it.

Despite all of this, there are benefits to recycling. With paper, for example, its recycling reduces the cutting down of trees, helping preserve habitats. It also doesn't end up in landfill, releasing methane. There is also a reduced water and chemical use. Paper is also generally recycled in the country it was used, e.g. the UK recycles its own paper waste.

In summary, recycling should occur where the other above options are exhausted. It may well be part of the transition to a better world, but it is a long way from the solution.

Rot; This is the composting of organic waste, such as from food or from the likes of grass cuttings, in order to create nutrient-rich compost to add to soil. Some supermarkets and restaurants also donate food to charity, or it is used for animal feed. Most of the rest should be composted but often it is not and will not occur unless legislation and significant consumer pressure is applied.

Not only is composting extremely beneficial, but it also keeps organic matter out of landfill, where it would decay without oxygen, releasing methane. At some landfill sites, tubes are sent deep inside to collect the methane. This is then burned for heating or electricity. The ideal, of course, is to switch to the aforementioned circular system, which has no need for these sites. Whilst we still have methane producing landfill, however, better that it is made use of, rather than allowing it to seep into the atmosphere. Having said that, it is estimated that there is currently so much organic matter within landfill in the UK that even if no more was added, they would continue to create methane for decades to come. Let's not add anymore then.

Summary

Material items are an important part of modern life. We have clearly seen above, however, that our constant demand for more and more items is having a significant environmental impact and is no way sustainable. We have also seen that there is another way; essentially, we need to switch from a linear to a circular model, where any waste just becomes the raw material for a new product.

Those who produce the products in the first place also need to be responsible for their entire environmental impact, from the mining of raw materials, all the way through to their eventual waste management. There also needs to be full transparency on this full cycle, so that consumers can make truly informed decisions.

We should also adjust our thinking and embrace each of the ten Rs, as well as make change in both how we produce and treat items, as well as the systems currently in place (more on this in Chapter Eight).

What we need now is the will, both as a society and as individuals, to carry out positive change. Remember, this is our climate, our future, and there is a better way.

CHAPTER 6

THE HOLY GRAIL OF TECHNOLOGY

There are many today who believe in so-called 'Ecological Modernism'. This is the unfettered belief that technology is the ultimate solution to our Environmental Crisis. After all, it permeates through our daily lives; technology is so intertwined in our lives that a new 'God' has spawned, in which our faith has been diverted towards technology and its ability to ever improve our lives. As Christians, including famously King Arthur, searched for the Holy Grail, so Ecological Modernists search for the next technology to save the world.

So, what is technology? By definition, it is the application of science to practical purposes, whether these be machines, devices, systems or methods. There is no doubt that we are living in a world in which we are ever more reliant upon technology. Consider how we rely upon it every day, with much of it never previously considered, let alone needed. Yet technology has transformed our lives, for both better and worse. Our dependence upon it has brought us both increasing comfort and opportunity, but it can also become addictive and take over our lives. How many of us would feel withdrawal after spending days, or even just hours, away from our phones? Technology has made many tasks easier, but at what cost?

Whilst we may have an ever-increasing number of 'friends' on social media, from around the world, how much has physical interaction been reduced? Indeed, many of us have become disconnected from the world around us, the people in our local communities, and from the intrinsic beauty of the natural world itself.

It is interesting that our faith in technology to save the world's ills, goes hand in hand with a denial that we will have to make actual changes within our lives. Our faith tells us that technology will solve it all or us. This is despite the fact that technology has had a turbulent history environmentally. Take the use of CFCs in deodorants and lead being added to petrol. Both caused significant harm, though thankfully, through science the issues were recognised and reported to such an extent that people stood up and took notice. Then with understanding and political will, adaptations were made to reduce and in fact, reverse the harm caused. This should indeed give us faith that when we have genuine comprehension, positive change can be made across the world. Our current Environmental Crisis, however, far outweighs anything we have seen before. Its scale demands a coming together of all of us.

Or maybe new technologies will be enough? In this chapter we consider their pros and cons, as well as the role they may have in reducing, and reversing, the harm we have caused. Let us now then look at a variety of different technologies to understand the difference that they can make.

Energy Resources

We know that the biggest contributor to Climate Change is the burning of fossil fuels. Within Chapter Three we saw the importance of technology as the world shifts from a reliance upon fossil fuels, towards the use of far greener energy resources. There is no need to repeat what we have already said, but technology plays, and will continue to have, a huge role in this transition. We also saw, however, that every energy resource comes with some environmental cost. Whilst some are far, far better than others, the time of using as much energy as we like is over. Along with changes in energy resources, we must change our habits.

This realisation is vital. Can technology help? Yes. Will it solve all of our issues by itself? Sadly not. Recognising this is key as we move forward towards a greener future. Yes, we want many of the comforts that modern life brings us, and indeed we want them for everyone, rather than the few, but it comes with a responsibility. We know that there are many around the world living far more difficult lives than ourselves. We must also recognise, however, that it is the wealthier amongst us who have created the most damage, whilst those less fortunate bear the brunt of our actions. Our responsibility then is to bring about positive change, not just for the benefit of others, but also ourselves. We will investigate this aspect further within Chapter Eight.

Carbon Capture and Storage

Nature is amazing at maintaining balance. It has, however, sprung forth a species so dominant that it is having an overwhelming effect on all others around the world. Humans. This includes taking carbon stored over millions of years in the past and burning it, releasing huge amounts into our atmosphere at unprecedented rates.

As mentioned in Chapter One, human activities have caused the concentration of carbon dioxide to increase from around two-hundred and eighty ppm to four-hundred and fifteen ppm. Even if we stopped the release of all human-caused Greenhouse gases right now then we still have a 'Legacy load', i.e., an amount that will cause the Earth's temperature to rise well into the future.

Clearly then, the amount of greenhouse gases in the atmosphere needs to decrease. There are some naturally occurring processes which cause this to happen. Plants, for example, are amazing at absorbing carbon dioxide from the atmosphere. Sadly, deforestation has led to a great reduction in the capacity of this natural solution. A proposed technological solution to this is 'Carbon Capture and Storage', or CCS.

There are two main forms of CCS. In the first, carbon dioxide is captured within the power station or large industry, and then stored before ever getting into the atmosphere. In the second, the so-called 'Direct Air Capture' or DAC, carbon dioxide from within the atmosphere is removed and stored. Let's then look at each of these two in turn.

Within power stations and industry, a CCS process is carried out either before or after combustion[1].

In 'Pre-combustion' CCS, the fossil fuel is reacted with oxygen, producing syngas, which is a mixture of hydrogen, carbon monoxide and carbon dioxide. The hydrogen is then used as the fuel, whether burned directly, or for use in hydrogen fuel cells. Water is added to the carbon monoxide, producing more carbon dioxide. All of this carbon dioxide is then removed for storage.

In 'Post-combustion' CCS, the fossil fuel is burned as normal (something wrong with the need to say that this is 'normal'), releasing carbon dioxide. The burning of fossil fuels releases a mixture of 'flue gases'. The carbon dioxide then must be removed from the others by using a combination of ammonia and steam. It can then be stored.

Another issue with the burning of fossil fuels is that carbon monoxide and soot are also produced. This occurs because there isn't enough oxygen within the furnace. Therefore, a process known as Oxy-combustion can also be used to cause complete combustion of the carbon, meaning that far less carbon monoxide and soot also produced. The extra carbon dioxide created can then be removed in post-combustion CCS as above.

In Direct Air Capture (DAC), there are two main systems used. In liquid DAC, air passes through liquid solutions, which removes the carbon dioxide, allowing the rest of the air back into the atmosphere. In solid DAC, substances in filters chemically combine with the carbon dioxide. The filters are later removed and heated, releasing the carbon dioxide for storage[2].

In other words, DAC is basically trying to mimic what plants do, as they remove carbon dioxide in the air through photosynthesis. Now, since carbon dioxide is still, thankfully, a very small proportion of the air, large volumes of air need to be in contact with into DAC systems for them to remove a suitable amount of carbon dioxide. This all requires significant amounts of energy to make the whole system effective.

In each of these, once the carbon dioxide has been captured, it then needs to be transported and stored safely. Since it is a gas, it would normally take up huge amount of space. The answer to this is to use more energy to compress it to a liquid, and then pump it deep underground. For example, to where fossil fuels have been removed from in the past. Oil companies have been flushing out oil to the surface for years by pumping carbon dioxide deep into the Earth, so this part is well understood. It does however need to be somewhere, like with the storage of nuclear waste, which is geologically stable so that the likes of earthquakes and underground water will not cause its release.

Another alternative is in pumping the carbon dioxide deep into the oceans. It is known that carbon dioxide is denser than sea water at large depths, so should stay down there. The issue, of course, is that carbon dioxide and water react, making the oceans more acidic. This will, as we have seen in other parts of the oceans, have a negative impact upon the species that live there.

Within all this there is an energy cost. Both pre- and post-combustion CCS make the whole process less efficient. DAC also requires energy. Compressing the gas to liquid does too.

Then there is the transport. The gas also needs to be pumped to where it is to be stored. Even then, best estimates suggest an underground storage time of around a thousand years. Given that this has, obviously, never actually been tried in practice, the truth may be far less time. Again, then we are leaving problems for future generations. But we can't carry on as we are.

We should also recognise that, whilst small scale CCS has been in place since 1972[3], none of this has been tested on a truly large scale. How will it all work? Even on smaller scale, the theoretical capturing of ninety percent of the carbon dioxide doesn't come close. Too many examples show that CCS isn't reliable and doesn't work anything like as effectively as reported[4]. Indeed, it has been found that some even emit more carbon dioxide than they capture[5]! This all shows that we cannot rely upon CCS. Future projections on its use rely upon technologies that are either not fully developed, or which haven't even been invented yet. One recently developed option involves bubbling carbon dioxide through a tube of a certain liquid metal[6]. This causes the carbon dioxide to be split up, with solid carbon flakes released, ready for collection. Whilst it hasn't yet been scaled up to industry levels, it certainly has potential. Probably the biggest advantage with this method is that it is solid carbon, rather than gaseous carbon dioxide which is stored. Not only can this carbon be turned into useful materials, but it means that it cannot escape back into the atmosphere if a leak occurs.

It should also be noted that technological CCS is expensive, which we ourselves pay for, either through taxation, or through increased fuel costs. Indeed, large scale CCS may even cause the cost of electricity from fossil fuels to double. This at a time when we are seeing ever decreasing prices of electricity from the likes of solar and wind. This gives us even more reason to reduce our use of fossil fuels.

One way to offset some of these costs is to make use of the captured gas. This captured carbon dioxide can be used for the likes of indoor plants, carbon fibre, fizzy drinks, and in the production of concrete and plastics. Another example is in feeding the carbon dioxide to algae which are later used as biofuel. This is referred to Carbon Capture, Storage and Utilisation (CCSU).

Two other ways in which the captured carbon can be made use of is in using basalt to turn the carbon into stone, and in turning it into diamonds. In the case of basalt, emitted carbon dioxide is combined with wastewater and literally injected into basalt well below the surface[7]. In this method the carbon becomes mineralised within two years. In the case of diamonds, carbon dioxide is combined with rainwater, heat, and pressure. Where wind and solar are used, these diamonds are made entirely from the sky[8]. In each of these cases there is far less of a storage issue and whilst they are not yet commercially viable, they certainly have potential.

There are, of course, better alternatives. We could not explore for, and exploit, any more fossil fuels sources in the first place. The carbon is safe there, underground, in the form of fossil fuels, and has been for millions of years. A pretty 'tried and tested' method from nature. Ok, so in the short-term we do still rely upon fossil fuels, but we need to move much faster to low-carbon solutions, many of which are discussed within this book.

But what about this 'Legacy load' of carbon already out there? Well, there are natural solutions to that too. Reforestation of land would bring huge benefits. Not only would a suitable combination of native trees capture and store huge amounts of carbon, but they would also provide habitats for so many other species, getting towards a balance in nature once more.

As discussed in Chapter Four, soil also stores huge amounts of carbon, with the likes of peat being particularly carbon rich. Not only do we need to protect these current natural carbon stores, but the recultivation of them also needs to happen. Our oceans also play a vital part in storing carbon. Plankton, kelp forests, salt marshes, mangroves, sea grass beds and coral reefs all play a vital role.

Overall, the best way to protect our natural stores of carbon is really to leave them alone. But we haven't done very well with that so some support will be needed to give this natural carbon sequestration a boost. Essentially, we should learn from and mimic nature as much as possible.

Having said all of this, the issues we are currently facing are vast. Fossil fuels are still used in huge quantities around the world, and not everywhere is able to move away from their use as fast as others. No single solution will solve it all. Alongside the natural solutions, technological CCS, including DAC, must play a role. In the future these will become more effective, but it may be many years until they can make a significant difference. As mentioned elsewhere though, every positive action is important and we must take every win, however seemingly small.

CCS must not, however, become an excuse to delay action and to continue to burn fossil fuels at alarming rates. Sadly, this very argument is being used by governments and the fossil fuel industry around the world, stating that we can continue to increase the burning of fossil fuels since the carbon can just be captured and stored afterwards. Not only do we know this is untrue, it goes against everything we need to be doing.

Whilst many individuals and small groups are working tirelessly, they can only do so much. Governments and corporations need to recognise and truly understand the magnitude of the issues. They must act together in reducing our reliance upon fossil fuels and implement both technological and natural forms of capturing and storing carbon. We will look more into putting pressure on such organisations in Chapter Eight.

Geoengineering

Geoengineering is another proposed technological solution to the world's climate issues. It is the deliberate large-scale intervention within the environment in order to counteract Global Warming and its effects.

There are several ways this in which this can happen. One, that in theory has potential, comes with huge risk. This is the use of stratospheric aerosols[9]. In other words, this is literally the release of large amounts of sun reflecting particles into our upper atmosphere. These would indeed reduce the amount of Sun's energy reaching the Earth's surface, and so reduce the temperature. But at what cost?

Firstly, not only would incoming heat be reduced from coming in by these particles, but also light. So, the Earth would be dimmer. It is suspected that it was the dimming of light due to huge amounts of dust being blown into our atmosphere sixty-five million years ago, which killed off the dinosaurs[10].

Also, we don't know the effect of such atmospheric particles on the likes of the ozone layer, weather, and other factors. It would be very tricky to remove these particles once they are there if it all went wrong. So, no, whilst I acknowledge there is some logic to this proposal, the risk seems far too great, particularly when there are so many other ways of dealing with our Environmental Crisis.

Despite this, scientists have released balloons, of the type that would carry the aerosols, into the stratosphere. Tests have been carried out on how to best maneuverer and position the balloons, as well as conducting further experiments on chemicals reactions which occur within that high up section of the atmosphere[11]. Whilst particulate matter has not yet been released, these experiments are part of paving the way for this to occur in the future.

Another method is the use of space reflectors[12]. These would, like the particles, reflect sunlight before it hits the Earth. This would be expensive and would require significant energy to install. It would also have a dimming effect. One advantage over the aerosols, however, is that they would be easier to remove once their job was done. It also wouldn't affect other aspects of our atmosphere. I would say that this one has potential, but we are some way off.

Another form of geoengineering is in altering the albedo of the planet so that the Earth's surface reflects more of the Sun's heat back into space[12]. The Earth's albedo is already changing, but in the wrong way. As ice melts, it reveals the darker surface below, which is much better at absorbing, rather than reflecting, heat. To counter this, living rooftops have been created. These both increase biodiversity and reduce air pollution. Other roofs, as well as roads, have been painted white. There is even a 'whitest paint ever', which can reflect just over ninety-eight percent of sunlight[13]. Whilst each occurrence may not seem like much, it all adds up and makes a difference. These methods can also be readjusted in the future if need be.

Of course, technological CCS, as mentioned previously, is also a form of geoengineering and we have noted the role it needs to play moving forward. But like CCS, geoengineering must not be used as a reason to not carry out all those positive changes in our lives which we know make a difference.

Transport

This is a pretty big section, so it seems sensible to split it into three separate forms; Land, Air and Sea.

Land transport

Before looking at motorised vehicles, let us first look at two low carbon, well-known, healthy modes of transport, namely walking and cycling. Now of course, journeys may be too long for these, and some people genuinely can't walk or cycle far at all, but let's embrace them when we have the opportunity.

One way to do this is in the design of cities centres and similar. When walking and cycling routes are useful, safe, comfortable, and interesting, then more people will take advantage of them. One step is to eliminate most motorised vehicles from such places (disabled and elderly may well need to continue to use them). The routes can be made aesthetically appealing, green, and diverse to appeal to potential users. This will lead to a happier, healthier population. There will be fewer accidents and lead to increased interaction between people and their environment.

There are around one-point-two-billion motor vehicles worldwide[14]. The fact that most of these run of fossil fuels, and so emit carbon dioxide into the atmosphere, is well documented. The bulk of these vehicles are cars, or motorbikes, and many of our journeys on such vehicles are for just one person. Surely there is a more efficient way to do things.

In many of the wealthier countries around the world, electric cars, and bikes, are purported to be the solution. There are indeed several advantages, environmentally, from using electric cars. Firstly, if the electricity hasn't been generated by burning fossil fuels elsewhere, then the carbon dioxide emissions will be reduced. This is even better if you are sure it has come from solar or wind, say from solar cells on the roof of your house.

Fossil fuel vehicles, especially those which run on diesel, also emit a significant amount of ground level fine particulate matter. This includes nitrogen oxide, black carbon and ozone[15]. This is a significant issue within large cities since long term exposure is linked to increased likelihood of asthma, emphysema (a lung condition that causes shortness of breath) and general lung decline. Indeed, increased ozone at ground level could have the same effect as smoking a packet of cigarettes every day[16]! Whilst this particulate matter is also produced at power stations, it is far easier to stop it from entering the atmosphere, and they tend not to be in the middle of large populations.

The emissions mentioned above are bad enough when they come from moving vehicles, but what about when such vehicles are left on unnecessarily? Idling is the act of leaving the engine on whilst stationary. This not only has fuel costs, but it also gives off the very same emissions mentioned above. And for what? Some may claim that turning vehicles on and off makes them less efficient overall, but, on average, this is only if they are turned on and off within around ten seconds[17]. Idling is bad enough anywhere, but it happens so much from cars, delivery vans, buses, and coaches in and around schools too. Young lungs are more susceptible to the effects of this pollution[18]. Even if someone doesn't want to act on Climate Change, surely no one wants to hurt young children.

Another advantage of electric vehicles, such as cars, lorries, and buses, is that they have reduced maintenance since they have around ten percent of the moving parts of other cars. They are also easier to construct, once you have all the materials, and cheaper to run. This is in large part because electric cars are more efficient. They are, however, well outside the price range of most people. In time, as production increases and technology improves, they will get cheaper, but for now, many people just can't afford one.

There is also a, mostly perceived, issue with range. Most journeys people take in their cars are to work, shops and leisure. Only rarely do most people go on longer journeys. Whilst there have been significant improvements in efficiency and range, electric cars do need recharging when travelling long distances. This also takes significantly longer than filling the car with fuel, and requires significant investment in charging infrastructure across the land.

For those who regularly drive long journeys, hybrid cars may be a suitable option until the range of electric vehicles increases. As noted in Chapter Three, hydrogen is also an option. This can be used as a fuel to generate electricity within the car. If the hydrogen is stored safely then it too can lead to significant positive impact, relative to fossil fuels.

A future alternative for electric vehicles is induction charging[19]. This requires significant input initially, but then, rather like how electricity is generated when a conductor passes through a magnetic field, the car will charge whilst driving along such inducting roads. The potential for this is amazing as it solves so many of the issues raised, but we are some way off this.

On top of cost and range, there are other issues, however. Currently lithium-ion batteries are used. These require several rare metals, such as lithium and cobalt. These metals need to be mined. This generally happens in parts of the world with limited regulation on workers' rights, health and safety and environmental impact.

For example, in some mines, acid baths are used, with the extracted material heated in a kiln. A tiny fraction of this is then removed as rare metals. The remainder, which is now toxic, is dumped back into the ground as waste. On top of this, in some mines around the world, such as in Congo, UNICEF estimates that forty thousand children, some as young as seven years old, are working in the cobalt mines[20]. And don't think that you haven't impacted on this if you don't have an electric car. The same is true for batteries in all our modern electric goods, such as mobile phones, tablets, and laptops. This may not be what you want to hear, but it doesn't stop it being true.

Another source of such rare metals is on the ocean floors. So-called deep-sea mining occurs in many locations around the world to remove such materials. This generally occurs between four and six kilometres below the surface. At such depths polymetallic nodules contain a high concentration of the precious metals which are in demand. The ecosystems at such depths are poorly understood, however, and the impact of such mining could be devastating[21].

Whilst deep sea mining already occurs below two-hundred metres, none currently occurs at the depths mentioned above. Licenses have been awarded for exploration though and some feel it is just a matter of time.

What is known is that when machines scrape the ocean floor, habitats are destroyed. This devastates many species not found anywhere else around the world. It can also create plumes of matter which could smother individuals and pose a particular threat to filter-feeding species such as Krill and Whale Sharks. The noise and light from such activities could also have a detrimental impact of sea-life.

Are these reasons enough to never use electric goods and vehicles? If you are in a position to be able to buy an electric car when a new car is needed then overall, despite the issues, an electric car is still better environmentally than one run on fossil fuels. If you are worried about those rare longer journeys, you could then hire a car or use public transport for those occasions. Otherwise, whilst personal choice must play a part, perhaps a better solution is both reduce how often we buy new electronic goods, recycle old ones so that their rare metals can be re-used, and put pressure on both governments and large corporations to act to ensure that this exploitation of both the environment and other human beings, ceases to exist. There is one traffic issue that electric cars don't solve however; there is too much on the roads. If you ever complain that you are stuck in traffic, it is worth remembering that you are traffic! Not only that, but there are significant emissions in the production of every vehicle. Indeed, the scrapping of a car in favour of a more environmentally friendly one should only be done if the car is hugely polluting or has come to a natural end.

Electric bikes, whilst only for individual use, do encourage a more regular use than human-powered bikes. They bring the added benefit of being great exercise, but can also be switched on when required, such as when going up-hill or tired. There is even research that suggests that they can be less polluting that 'push bikes', particularly if the energy the person uses is from high emission foods such as beef[22].

So, how do we reduce traffic on the road, demand for new vehicles, polluting emissions, whilst maintaining convenience at a reasonable cost? The simple answer is to have more people per vehicle. For example, a car-pooling scheme could be used. Not only does this reduce emissions and costs, but it has the added benefit of creating extra social interaction in person, something that has, to our detriment, reduced in modern times. It may be that others you already know take similar, or even the same, journeys. There are also several car-pooling apps out there for people to use.

An even better solution is to have a regular, reliable, affordable, comfortable public transport operation in place. This could use a system which makes it easy to pay for too. Covid-19 may have reduced the desire of some people to be with many others, but if managed suitably then there are many wins from this. Let's face it, cars are the ultimate depreciating purchase. As soon as you drive it off the forecourt, its price starts going down.

Public transport can bring greater transport efficiency, lower costs, and less pollution, especially if these public vehicles were electric. There would also be reduced traffic jams and fewer accidents. It is worth noting here that whilst the emissions from a single bus will be greater than a single car, you wouldn't need many people on the bus for the emissions per person to be way less than that of a car[23].

Another significant benefit of public transport is that it reduces inequality as it brings people from all walks of life together. Where we all use the same transport there becomes less of a 'them and us'. This bringing together, including with those who can't drive, whether due to age, disability or other, provides us with shared experiences, which is so vital in creating a harmonious society. Prejudice and intolerance come from ignorance. The more people mix, the further these barriers will break down.

Increasing cycle lanes should also occur, especially in cities. Not only are they safer for cyclists, but they also encourage extra use, and thus reduce the number of cars on our roads. This makes transport more efficient for everyone. It also, therefore, leads to fewer emissions, and a healthier population. Win, win!

So why isn't all of this happening already? The car and fossil fuel industries have vast power and influence. For generations we have been bombarded with the message that to be successful, we need more. To be free we need our own car. It's even explicit in some adverts still today. But these industries don't just hold sway over individuals, they also do so with governments.

These governments may also be reluctant to pay for, with our money, the high initial costs in providing the infrastructure. It is also true that there will be greenhouse gas emissions in the initial construction of such a system, but the benefits both to society and to the environment, will soon outweigh this initial outlay. It should also be noted that some of these initial costs can come from money assigned to creating new roads, which won't be needed as there will be less traffic on the roads.

High speed electric railways, for example, are smooth, the fastest way to travel over a few hundred miles, and have far lower carbon emissions than flying or by individuals driving[23]. They also have the added benefit that railway stations can be within a city centre, whereas airports cannot. It is important here though, as the UK has seen with HS2, that any new railway lines be chosen carefully as to not obliterate the likes of ancient woodlands. This rail should also be electrified as to reduce the need for diesel.

Air transport

Perhaps the mode of transport that receives the most negative press is flying. This is well deserved. As we know, the burning of fuels releases carbon dioxide into the atmosphere. In these terms alone, flying can have fewer emissions than one or two other modes of transport. So why the bad press? Well, it's because that is not the whole picture. Plane fuel also releases other greenhouse gases such as nitrogen oxides. Where the emissions are released also plays a role.

When emissions occur high up in the atmosphere, they have a greater impact on Global Warming.

There are also the contrails. These condensation trails are line-shaped water clouds, often in the form of ice crystals. As we know, water is a powerful greenhouse gas. When these contrails are produced high in the atmosphere, they reflect heat back down to Earth, increasing warming even further[24]. Time of day also affects how much warming is caused by these contrails. In the day, contrails reflect heat back to the Earth, but they also reflect radiation from the Sun back into space. At night-time, contrails still reflect heat back to the Earth, but now don't reflect solar light and heat, since that side of the Earth is facing away from the Sun. This means that night flights lead to greater warming than flights during the day[24].

Whilst the amount of warming caused directly from aviation is still relatively small compared to other industries, with increasing flights, and increased green initiatives around the world, the proportion which is due to flying is increasing. Therefore, people can be seen protesting the expansion of airports. Larger airports mean more flights, which means greater emissions, leading to greater warming. Airport expansion also increases trade, encouraging people to buy more and more from around the world. This is a trend that cannot continue if we are to truly mitigate our Environmental Crisis. More on this in Chapter Eight.

A question we need to ask ourselves too is whether we really need to fly so often. Even in a relatively wealthy country like the UK, only fifteen percent of adults take seventy percent of flights[25]. If we also consider less affluent countries, where the percentage who fly will be far lower, then it is sensible to suggest that only a small percentage of the world's population take most of its flights. Can this be justified? Many of these, of course, are business flights. For all its negatives, one positive from Covid-19 is that many businesses now recognise that there is often no need to jet set around the world when a video call can link those across the world from their own offices. Let's see what happens as the we move forward.

So how does technology help here? Well, the design of both the engines and the rest of the plane, has led to increased efficiency. This is good. But not enough. Some flights, well, just a few, now use biofuels, or at least a mixture of fossil fuels and biofuels. This is preferred to fossil fuels alone, but as we saw in Chapter Three, is not a true solution. So, what about electric planes? Well, current batteries cannot hold the energy required for such long journeys. The number of batteries required for long plane journey would make it very heavy and take up too much space. Until we find a way to produce light, high energy density batteries, which don't cause human and environmental suffering, then we really need to significantly reduce the number of flights that we are taking.

I know that some people continue to fly regularly, whilst offsetting the emissions from their flights, but this is a bit of a misnomer. At the costs they are sold for, and at what they advertise to do, the whole enhanced greenhouse effect could be eradicated by far less than one percent of the world's GDP. If only this was true. Sadly, of course, it isn't. OK, paying for trees to be planted or solar cells to be implemented around the world is great, but these are things that need to be happening anyway. Also, solar cells don't remove carbon from the atmosphere, they just don't add as much as fossil fuels. The impact of offsetting is notoriously difficult to calculate, and it may well do little more than reduce your guilt. If you do choose to take a flight then offsetting is probably better than doing nothing at all, but it won't become a net zero flight, no matter how much you try.

Just an addition. As mentioned in the introduction, I have a degree in Astrophysics. I love space and find it, and the exploration of our Universe fascinating. But billionaires funding space tourism just to satisfy their own egos, has got to be one of the worse environmental actions of modern times[26]. I have made a point of not judging others within this book, and I believe with all my heart that every single one of us has an important role to play in making our world a better place, but just 'No' on this one.

Sea transport

Ships play a vital part in our global trade, indeed, most of it is moved around the world via shipping. Not only is it significantly cheaper than by air, but air freight (the transport of goods by air) emits up to fifty times more emissions than ship freight (the transport of goods via sea)[27]. Still, since so much ship freighting occurs, the total emissions do have an impact and so reducing them is important.

Many ships have traditionally run off cheap, low-grade fuel. The burning of this fuel not only emits carbon dioxide, but also a high proportion of particulate matter. A number of these cargo ships are now switching to diesel instead. I know! When diesel is the cleaner fuel that what the heck was the previous one like? But any positive change makes a difference.

Due to the long distances involved, ships, like planes, are some way off becoming electric. For now, we can reduce our consumption of goods, leading to less need for cargo ships, and acknowledge that it is improving, and way better than transport via planes.

Digital Revolution

The digital revolution, which began in the late twentieth century, has changed our world beyond recognition. Whilst so many of us depend upon it in our lives, there are clear pros, but also significant cons, of such a drastic change.

Many of us are so caught up in the digital world that we hardly notice. This includes the internet, social media, smart phones, streaming shows and music on demand, and instant communication wherever you are in the world. It is now an intrinsic part of so many of our lives.

In many ways this has brought us together. We can now chat to friends and colleagues, like-minded individuals, and those of radically different views, wherever and whenever we like. And in an instant. It has also led to greater transparency, created jobs, and led to a greater sharing of expertise in the likes of science and medicine.

This does come with a darker side, however. Social media forces us to compare ourselves to the current ideals of the day, whether physically, financially, intellectually, or spiritually. This has increased our own feelings of insecurity, especially given that so much online is effectively a lie, with photoshop, specialised lighting, camera angles, and many people only sharing what they perceive to be their best. This makes the lives of others seem so much more interesting than our own. It has also taken 'fake news', racism and all other forms of intolerance to a whole new level, where trolls hide behind their keyboards, safe in their anonymity, saying whatever they like to make their own lives feel better. So much of this they would never say in person. Alongside this, face to face contact has reduced significantly[28], with people more interested in what is happening in the fake world than in the real. The social pros and cons of the digital revolution cannot be underestimated. Whilst it has brought some people together, divisions have also been magnified. There has been, and continues to be, an alarming rise in mental health issues, especially within young people[29].

The issues mentioned above have been explored in far more detail elsewhere, but one thing to reiterate is that the Environmental Crisis is very real and requires a global commitment. Whilst this digital revolution has the potential to bring us together for such action, that is not what is happening. There is too much noise, too many distractions, and too much spreading of doubt.

There is still hope, however. And hope or not, we will continue to fight for a better future for us all.

So, what about the environmental impact of this all? Well, as mentioned above, since video calls and the like are now instant and across the world, there is no need for so many business flights. This is clearly a positive. The big digital companies themselves, such as Google, Apple, Facebook, and Microsoft, have enormous digital centres, where they store their data. These use a vast amount of electricity. Indeed, around two and a half percent of all emissions come from the use of ICT[30]. This may not sound like much, but if it was a country, this would put it in sixth place.

Saying that, there is much discussion about whether the use of ICT actually has a net negative, rather than positive carbon footprint, given that it makes so many things more efficient. Whilst this is true, we are also doing many things that we wouldn't previously have done at all. This means that the overall usage of ICT has increased, leading to far greater emissions from digital technology than ever before.

The largest use of the internet is in messaging, particularly the use of email. Around three-hundred-billion emails are sent each day, to around four billion users. This averages at around seventy-five emails every single day, with about half of them being spam! Thankfully, most of us never even see this spam, due to the employment of filters, and simply delete any spam that does get through. Whilst potentially annoying, it doesn't contribute as much as when you send an email and copy in more people than necessary.

When we do this, we take up people's time, increase the carbon footprint of the email, and model behaviour which encourages others to do similar. Whilst each individual email has a far lower carbon footprint than a 'snail-mail' letter, no one ever wrote this many letters every day. Here we have yet another example of when technology makes things more efficient, it leads to an overall increase in carbon emissions. A change of culture is needed. The number of emails we receive each day is not sustainable, from either an environmental or time perspective.

We know that the big tech companies are extremely skilful of 'sucking us in' and ensuring that we spend as much time as possible on their platforms. The more time we spend there, the greater their profits. But surely they are doing their part? Well, there is no doubt that some are better than others. Microsoft has committed to be carbon negative by 2030. They have also stated that by 2050 they will have removed all carbon they have ever emitted into the atmosphere[31]. This does rely upon CCS, but at least they seem to be moving in the right direction.

Amazon has committed to being carbon net zero by 2040. They do, of course, have the added reliance upon fossil fuel-based delivery of their goods. They have also committed to one hundred percent renewable energy in their data centres[32]. They are not there yet, and we know that some renewables are much better than others. But it is still better that fossil fuels. As part of this they themselves have a vast array of solar cells as well as wind farms. It is unclear how the use of their own cargo planes, namely Amazon Air, to get products to where they need to be a day earlier, fits in with this. They have even had their own workers protest their carbon emissions.

Google has used one hundred percent renewable energy since 2017. Again, not perfect since they still use so much, but better than fossil fuels. They are also working hard on increasing efficiency. For example, they have reported that between 2010 and 2018 their computing power increased by five-hundred and fifty percent, but the energy use in their data centres 'only' increased by six percent. Google also has its own solar farms[33]. In many ways they, and others, have had to become so much better. With the likes of search engines like Ecosia planting trees for searches[34] (that click on advertising), Google need to be seen to be doing their part or risk losing customers.

There is no doubt then that the big tech companies are at least being seen to make improvements. They still have a long way to go but the digital world is now so ingrained in modern life that most of us will continue to embrace it. We must, however, encourage, and apply pressure, for them to use their vast wealth and influence to make a significantly greater positive impact.

One area of the digital world which is extremely energy intensive however is Cryptocurrency. Cryptocurrency is a digital currency which can be used to buy online goods and services. Whilst it has grown in popularity, a relatively small percentage of people use it, yet it still leads to more emissions each year than a million transatlantic flights[35]!

Medicine

Medicine is big business. Indeed, it is one of the largest industries in the world, including many pharmaceutical corporations. The pharmaceutical industry is the discovery, development, production, and marketing of drugs. Many people rely upon the medicine that they take to keep them healthy and pretty much all of us have taken medicine at some point or another in our lives.

One aspect of the pharmaceutical industry that is less well known is that its greenhouse gas emissions are significantly greater than all vehicles on the road[36]. At each stage of the 'discovery, development, production and marketing' of drugs, energy is required. Whilst it is not reported in the media as much as with other industries, these companies know their emissions are too high, so need to take appropriate steps to reduce their emissions now.

Another major issue driven by the pharmaceutical industry in the contamination of land and water with waste products. This is particularly prevalent in areas near manufacturing facilities and the associated wastewater treatment plants. This includes the release of hormones into the environment which are then taken in by a variety of species. For example, many male fish in Europe's rivers are now considered 'intersex' as they also display several female characteristics. This is likely due to the absorption of female hormones released into these rivers, such as from the contraceptive pill[37].

Another significant issue caused directly by the pharmaceutical industry is that, as noted in Chapter Four, its overuse leads to the creation of bacteria which are resistant to antibiotics. This had already led to a number of so-called 'superbugs' such as MRSA. The development of antibiotics had brought huge health benefits, but if we get to the stage where enough bacteria are resistant to these drugs then we will no longer be able to treat them effectively and the cost will be our health. In India, for example, an estimated fifty-eight-thousand new-born babies die from drug resistant infections every year[38]! So, whilst it is important to continue the use of antibiotics, they should only be used when necessary. Antibiotics are also used far more with livestock than they are with humans, often unnecessarily. When antibiotics, antifungals and similar are released into our waterways, they also lead to drug resistant bacteria and fungi. Not good!

This is all clearly unacceptable and the influence that 'big pharma' has over governments needs to reduce. Then, proper legislation, which is monitored and enforced, needs to be in place to protect us all. Clearly leaving this responsibility to rely upon the industry's good nature is not working.

Lighting and Other Appliances

Lights at night-time have been a vital part of human progress and survival for thousands of years. It started with fires to keep us warm and predators away. The first lamps were then invented some seventy thousand years ago[39], with candles coming in around three thousand BCE. Lamps then developed, with the use of kerosene, and then electricity, within the nineteenth century.

Since that time, the electric lamp has gone through many transformations, including filament lamps, energy saving lamps and now as LEDs. Through this, the lamps have improved in both safety and efficiency.

Filament lamps only started to be phased out of use within the twenty-first century. These are around ten percent efficient. This means that only ten percent of the energy is converted into light, whereas the rest becomes heat. Not only does that make these lamps dangerously hot, but most of the energy is wasted. Ninety percent of the electricity going into these bulbs isn't used for its designed purpose!

There is no doubt that the invention of LEDs has had a significant impact. These are more like fifty percent efficient, meaning they are far safer, since they don't get as hot, and require far less electricity going into them. They also last a lot, lot longer. So, whilst an individual LED may be more expensive to buy in the first place, not only are they far better environmentally, but they are also much cheaper in the long run. Win, win!

Another beauty of LEDs is that they have such low energy requirements, that they can be used in places around the world which are not connected to an electrical grid. Small solar cells can be used instead, which replaces the use of kerosene lamps in poorer and remote locations. Note: kerosene lamps are not only a fire hazard, but they also emit greenhouse gases, as well as particulate matter, which damages the lungs of those who use them, often mothers with young children!

Another lighting invention is the gravity light. This can be used anywhere around the world and has no need for burning or an external electrical supply. The gravity light is simply attached to the ceiling. It contains a bag which is then filled with sand and rocks. The bag is then lifted, and as it slowly descends, it drives a generator, producing enough energy to light the LED[40]. Ok, so you must lift the bag again every twenty minutes or so, but the importance of such an invention to the lives of people around the word should not be underestimated. A newer extension upon this is the Nowlight, in which you pull a cord, rather than lifting a bag. In these, just one minute of pulling on the cord can generate enough electricity for about two hours[40]!

Similar improvements in efficiencies can be seen across many other appliances. Whilst this is a positive, we also need a change in culture. The issue is that the increase in efficiency can cause people to think that it is ok to leave such appliances running. Of course, if we leave it on for longer then this can not only reduce the gain, but can even cause increased costs, energy use and emissions.

Dealing with Plastic

As we saw in Chapter Five, the world has a significant plastic problem which doesn't look like it's going away anytime soon. Part of the issue is that plastics are so difficult to genuinely break down into raw materials.

There are, however, a couple of possibilities using biological organisms to do so.

One method was discovered by scientists outside a recycling factory in Japan. A species of bacteria was found to have evolved to decompose a certain type of plastic[41]. To do this, it developed an enzyme which is capable of breaking down the plastic molecules. Whilst the whole scale release of these bacteria is clearly not something that anyone is considering, there is potential here for scientists to extract and bioengineer this enzyme. These enzymes could then be used upon plastic in large facilities, or even sprayed onto waste that has accumulated within our environment. Obviously, studies will need to be carried out to ensure it doesn't do more harm than good, but there is certainly potential here.

The other biological option in dealing with plastics, is with the use of certain types of mushroom[42]. For example, a rare mushroom from the Amazon rainforest was discovered to digest and break down one type of plastic. Another was found to 'eat' plastic at a waste disposal site in Pakistan. More research needs to be done here, but it certainly looks like there is potential to make a difference in the future.

Having said all of this, a far better answer is to remove our reliance upon these plastics. We must also recognise, however, that since there is so much plastic already out there, technology may well play an important role in its removal from our environment.

Summary

Our use of technology has brought us so many benefits, but it has come at a cost. Technology is rightly here to stay. It is a vital part of our lives, but the associated corporations, with government support, need to take a more active role in improving their environmental impact. There are many technologies that are already making a positive difference. We also have the even better 'nature's technology' which has led to a healthy planet for most of its existence. Instead of waiting on technologies that have not even been invented, we have the solutions now, both natural and artificial. The time to act is now! It is indeed true that some technological advances will make improvements as we move forward, but we know how to mitigate the Environmental Crisis without them. To not do so is surmountable to maintaining the current status quo which is doing so much harm to our planet. The two main dangers here are Ecological Modernism and certain types of Geoengineering. It is political will, not further advances in technology, that is needed. Remember, this is our climate, our future, and there is a better way.

CHAPTER 7

BIODIVERSITY; ISN'T NATURE WONDERFUL

We currently know of only one planet in the entire Universe on which life exists, Earth. It began over three and a half billion years ago[1], less than a billion years after the formation of the Earth itself. The presence of living organisms has not only changed the makeup of our planet but has created the wonderous nature that we see all around us, including our very selves.

As life has evolved and adapted over time, a key component of its survival is its vast and diverse variation, known as biodiversity. Biodiversity is literally biological diversity. It is the variety of species, and variation within each species, both as a whole and within individual ecosystems.

We also know that biodiversity on Earth is under threat from our very actions. Climate Change, pollution, habitat destruction, invasive species, and over-harvesting such as overfishing, of our biological resources, threaten life itself. Indeed, human-caused extinction is now at such a rate that many scientists are stating that we are in a sixth mass extinction[2].

The inherent beauty of nature and the awe and wonder it brings to our lives are not the only reasons why it is vital to maintain biodiversity on Earth. Our future existence as a species depends upon it.

Extinctions

Extinction is a natural part of life. Throughout the history of life on Earth, the vast majority of species that have existed are now extinct. Normally when extinctions occur, the void is filled as other species adapt. There are also, however, so-called mass extinction events.

Mass extinctions occur when species vanish much faster than they can be replaced. This occurs when at least fifty percent of the species on Earth die out within a relatively short time frame. There have been five mass extinctions in the past, ranging from fifty percent to ninety-six percent of all species dying out. These have been caused by a range of events including volcanic activity, continental drift, Climate Change, reductions in oxygen, and famously, as in the time of the time of the dinosaurs sixty-five million years ago, when an asteroid hit the Earth[2].

These mass extinctions, as well as the 'background' extinction rate, can be determined from fossil records alongside a variety of methods to determine dates, atmospheric composition, and temperatures.

The current rate of extinction is estimated to be over one thousand times the background extinction rate. That is, species are dying out at a rate over one thousand times higher than they would be if it were not for the action of humans[3]. We know what is causing such a high extinction rate. We could list a whole variety of reasons, but essentially there is one overlying cause, us.

Examples of human-caused extinctions include the Passenger Pigeon from North America, the Dodo from Mauritius, the Golden Toad (Zapo Dorado) from Costa Rica and the Pyrenean Ibex from Europe[4].

Before we look into the importance of reducing extinctions and conserving biodiversity, just consider how it makes you feel that we, as humans, have directly caused entire species to be wiped from the face of the Earth, never to be seen again. How do you feel about the possibility that your children will grow up in a world without such majestic creatures as tigers, Leatherback Turtles, Polar Bears, and Golden Eagles? Let us consider whether we have the right to not only wipe species from existence, but to deprive future generations from ever seeing them.

Why Conserve Biodiversity

So, there is clearly a biodiversity crisis, but what other reasons are there for wanting to conserve such variety of life?

The key point here is the fact that all life is intertwined. Whilst we often study individual ecosystems, the truth is that all life is linked together and that when one species dies out then many others are impacted too. This includes so-called keystone species. These are the ones who have the greatest impact upon those around them. It is basically this species that helps that ecosystem sustain itself. The loss of a keystone species can eventually cause the complete collapse of that community. Part of the issue here though is that we often don't even know which are these keystone species. They could be any from top predators, such as wolves, to maple trees, coral, hummingbirds, and mangroves.

Also, in times of pests, disease, drought, floods, cyclones, and extremes of temperature, all of which are becoming more prevalent, the greater the biodiversity, the greater the likelihood that some will survive. If species numbers are limited, or genetically similar, then they are all likely to be affected in the same way. For example, compared to having a rich biodiversity, if all plants in an area are of the same species, then there is a greater chance of them all dying out. Essentially, the greater the biodiversity, the higher the chances of survival. It's as if nature got it right.

But there are some people for whom the reasons laid out above are not sufficient, so let's look at ways in which having a rich biodiversity also improves human lives.

Humans use many species as resources. These include wood (for buildings, tools, furniture and fuel), fibres (e.g. cotton, wool, paper, silk and rayon), oils (for cooking, foods, lubricants and soaps), and other fuels (such as charcoal and alcohol).

We also rely upon biodiversity for food. We find that the variety of crops and livestock species that most people eat is actually relatively small. Therefore, to increase our own food security, new foods from nature are required. These include varieties such as drought-resistant Morama beans from Africa, perennial maize from Mexico, American Bisons, and Giant Land Snails. Whilst some of these may not salivate your taste buds, increased food security through biodiversity should not be underestimated.

We are also inspired by the natural world in our designs. Biomimetics is the imitation of nature within technology to improve human lives. Many species have highly adapted traits which have had millions of years of evolutionary pressure to produce them. As humans we can copy and make use of these. Examples include the shape of hawks (for fighter jets), shark skin (for ship surfaces and hospital operating rooms), termite mounds and their natural solar driven ventilation system (for large buildings), seeds with burrs (for Velcro), lotus leaves (for self-cleaning glass), spider silk for air bags and armour, and bats and dolphins for ultrasound imaging. Nature truly is amazing!

Many plants and animals have chemicals which protect them both from disease and being eaten. These can be adapted to create medicines. This includes from poppies, willow trees, sea sponges, yams, and shark blood. If we allow extinctions to continue at such a high rate, we may inadvertently cause a species to die out that contained the very chemical we need to cure new diseases, or even ones we are more familiar with, like cancer. It could be your loved ones, or even yourself, who would have benefitted from this. Is it really worth letting these species go extinct?

In Chapter Four we saw many issues relating to pesticides. Instead, we can use natural pest control to protect out crops. There are many examples in use today, including the cultivation of beetle banks to support predators such as ladybirds.

Wild species can also be used as a genetic resource. Our crops and livestock are fairly limited in biodiversity, but similar species can be found in nature which are more genetically diverse. Crossbreeding can cause commercial species to become better at fighting off disease, more drought resistant, more suited to growing in difficult conditions and also improve both yield and taste. This further increases our food security.

Hopefully now we have a clear understanding of the need to conserve biodiversity. Let us now then look at a variety of ecosystems around the world to better comprehend the issues.

Oceans

The oceans and seas of our planet contain a vast variety of species, from the Blue Whale; the largest animal that has ever lived, to the multitude of microscopic phytoplankton which underpin so much life.

Since the beginning of the industrial revolution, over thirty percent of the carbon dioxide emitted by human activities has been absorbed by the oceans[5]. They have also absorbed around ninety percent of the extra heat from the enhanced greenhouse effect[6]. Without these two processes occurring, the atmosphere would contain significantly more greenhouse gases, and would be a lot warmer. But what effect is this having on the oceans themselves?

Well, within Chapter One we saw that increasing temperatures can lead to changes in ocean currents. It also leads to melting ice and thermal expansion, leading to sea levels rising which can devastate coastal communities. Warmer water is also less able to contain dissolved oxygen, reducing how much of this life-giving element is available for ocean species. The extra energy in the waters can also fuel cyclones and heavy rainfall, increasing their severity. The absorption of carbon dioxide also makes the oceans more acidic[7] since the carbon dioxide literally reacts with the water, producing carbonic acid. But what affect is this having on marine life?

One well known ocean habitat is coral reefs. Coral is made up of millions of tiny polyps. These remove calcium from the water to produce hard lard layers of calcium carbonate, which serve to protect them. The polyps also have a symbiotic, or mutually beneficial, relationship with tiny algae, which gives the coral its amazing colours. These Coral reefs are sometimes referred to as 'the rainforests of the oceans' since they have such a high biodiversity, containing up to twenty-five percent of all marine species[8].

These reefs, however, are extremely sensitive to rising sea temperatures, pollution, and ocean acidification. This can cause the algae to be expelled from the coral, giving it a white colour. This is known as coral bleaching. Whilst not dead at this stage, it is a clear sign the things are not as they should be and if conditions do not improve then the coral will die. Since coral is also a keystone species, this will have a significant impact on the rest of reef life.

Before the 1980s, coral bleaching had only been recorded after a hurricane[9]. Since then, major bleaching events have followed with increasing severity and regularity at coral reefs all over the world. It is now estimated that twenty-seven percent of all coral reefs have been destroyed, with another thirty percent in grave danger[10]. If global temperatures increase by two degrees Celsius above pre-industrial times, then it is possible that all reefs could be lost. New ones could be formed but this process takes thousands of years and assumes that some of the polyps survive to do so. Healthy coral reefs are vital for so many species, including fish, lobsters, crabs, clams, octopuses, sea turtles and sponges. They are also key ecosystems which support millions of humans across the world. Coral reefs provide people with jobs in food, recreation, and tourism. They are also a spawning ground and nursery for many fish which then move elsewhere. Reefs are estimated to contribute billions of pounds to world economies and have also been an extremely useful resource in developing new medicines.

One example is the Great Barrier Reef off the coast of Australia. This is the largest living structure on Earth. It is over two thousand three hundred kilometres long and is home to over nine thousand species[11]. It has also existed for over forty million years. A recent study showed that the number of corals on the Great Barrier Reef had reduced by over fifty percent over the last three decades[12]. This is an extraordinary number.

This has caused a large array of species to become threatened with extinction. For example, the Short-Nosed Sea Snake[13] hasn't been seen since 2000 and may well be extinct already. Other reef species which are threatened include the Ridley Sea Turtle, the Great Hammerhead Shark, the Nassau Grouper and the Japanese Eel[14]. None of this needs to happen, but it is. But all is not lost. There are organisations and groups of people working hard to restore the reefs. This includes the growing and planting of healthy corals, the removal of invasive species, and responding to emergencies[15]. This is already making an impact but more needs to be done.

Another vital marine habitat under threat are mangroves. These are a range of tropical marine trees and shrubs which have prop roots (i.e. roots which help prop up the plant). They form dense tangles of plant material along a coastline. These mangroves are not only important carbon sinks, but they protect coastal land from erosion and the effects of storms and sea-level rise. In fact, these natural defences are noticeably more effective than man-made barriers. Yet another example where nature got it right.

Mangrove forests also support unique ecosystems which are full of biodiversity. They also provide us with food and tourism. Despite this, studies suggest a loss of almost ten-thousand square kilometres since 1996[16]. This is an area of similar size to the whole of The Gambia or The Bahamas! For those in the UK, this is larger than Wiltshire, Somerset and Berkshire combined.

It may be a surprise to learn that so much of our mangroves are being lost considering they are not as sensitive to changes in temperature, pollution or acidification as coral reefs. So, what is causing it? Well, it's us again. Mangrove forests are being deliberately cut down and removed to make way for housing, agriculture, such as palm oil plantations, and aquaculture.

In Thailand, for example, around sixty-seven percent of all mangroves have been lost[17]. The main cause of this is the replacement of mangroves by shrimp farms, to feed people around the world[18]. So don't go thinking this is a Thai issue, we are all involved.

The loss of mangroves around the world has led to many species being threatened with extinction. These include the Dugong, the Pygmy Three-Toed Sloth, the

Bengal Tiger, the Mangrove Hummingbird, the King Cobra, and the Sharptooth Lemon Shark, as well as the mangrove plants themselves[19].

Like with coral reefs, efforts are being made to restore our mangroves. Whilst this is important, we could just stop destroying these habitats in the first place.

Perhaps a more publicised issue concerning our oceans is overfishing. Programmes such as Seapiracy[20] have done much to raise awareness on this. Whilst such shows may make the odd exaggeration or quote out of context, their overwhelming message is correct; we are overfishing our oceans to the extent that many fish populations around the world are in danger of collapse. Our oceans were once thought of as an infinite supply of fresh, wholesome seafood. Not anymore. Not only have we seen that plastic pollution has meant that seafood generally contains high quantities of microplastics[21], but our consumption of fish and other marine animals has led to such reductions in populations that fishing itself is no longer viable in many places.

It is estimated that the number of fish in our oceans has halved since 1970![22]. Some species, such as tuna, mackerel and bonito have declined by almost seventy-five percent. How can this be sustainable? Since 1984, the population of Atlantic Cod has decreased by around seventeen percent every single year[23]. This favourite of British 'Fish & Chips' is in serious decline. The solution seems easy, don't eat seafood. Yet millions of people around the world rely upon these fish to survive. So, what do we do? Well, as humans we are not all in the same situation. There are people around the world who won't manage without consuming some fish, but this is not what is causing overfishing.

We now have a society, particularly in the wealthier countries, where everything is available all the time, as we demand ever more. It is not that the human population has doubled over the last fifty years, our fish consumption per person has doubled[24]. This means that we are removing four times as many fish from the oceans as we were in 1970. Modern fishing methods also produce huge amounts of waste, leading to even more fish dying than need be.

So, in recognising the fact that some people rely upon seafood for their health, whilst also being clear that current levels of fishing is unsustainable, what should the rest of us do? Well, we can be the ones who reduce consumption. Indeed, many people now refuse to eat fish, whilst others have reduced significantly. What will you do?

Sadly, the situation with many marine reptiles, mammals and sea birds is similar, with estimates suggesting the number of these halving since 1970[25]. Sea turtles, for example, who have existed for over one hundred million years, are now in significant decline. Each species is now either vulnerable or endangered. One example is that in the East Pacific, the majestic Leatherback Turtle has plummeted in number by a massive ninety percent in just thirty years[26]. There are many turtle conservation projects around the world, including those protecting their eggs from poaching, but far too many still end up being eaten, illegally traded, or as bycatch of the fishing industry.

If this wasn't enough, the fishing industry has a disturbing relationship with modern slavery[27]. Most of us would be horrified to think that slavery has played a part in delivering the food to our plates, but that is exactly what is happening. And whilst some may point the finger at other countries, consider this; if we are eating that produce then we are part of the problem, and there have been media reports of modern slavery and labour abuses aboard vessels from the likes of the UK and the USA, as well as Asia. Is this something you want to support?

Thankfully, it is not all doom and gloom. There are success stories too. It is estimated that in the early twentieth century, whale hunting was so extreme that there were just a few hundred humpback whales remaining[27]. Then in the mid-twentieth century the International Whaling Convention was signed banning the commercial hunting of these exceptional animals[28]. Some 'scientific' whaling continued, and some countries still hunt whales today[29], but there is no doubt that the situation has improved greatly, and current numbers have risen to pre-hunting days.

Not only are whales a vital part of the ocean ecosystem, but they also play a, perhaps surprising role, in carbon sequestration[30]. Phytoplankton need both sunlight and nutrients. One issue is that nutrients can sink, leaving the surface of the ocean nutrient deficient. This is where whales come in. Their release of faeces provides these vital nutrients. Since they eat at depth but defecate near the surface, they are essentially acting as a nutrient transmitter from one place to another. Their movement also spreads and mixes these nutrients into the water, thus allowing for more phytoplankton, and thus storing more carbon. Amazing! Not only that, but whales are huge and store significant amounts of carbon themselves. When they die, they sink, taking their carbon with them, to where it can enter sediments and be stored for large periods of time.

Another area in which we can see success is with kelp forests. Kelp is a type of seaweed and has a large variety of species. The forests they form cover an area of three-hundred-and-sixty million acres around the world[31]. And are amazing! They support a high level of biodiversity and grow faster than trees and bamboo (which is a grass). The protection and creation of kelp forests would bring life back to large ocean areas. They can also help provide us with food, fertiliser, fibres and biofuels[32]. Kelp also absorbs and stores large amounts of carbon. And they are suitable for so many locations. By using structures and attaching the kelp, you can create floating forests which also don't interfere with shipping.

Another example is with seagrass. This amazing plant captures carbon up to thirty-five times faster than tropical rain forests and is responsible for absorbing ten percent of the ocean's carbon, even though it covers only zero-point-two percent of the ocean floor[33]. Like kelp and coral reefs, seagrass plays a huge role in supporting biodiversity, with a ten thousand square metre area being able to maintain eight thousand fish and over a million invertebrates. It is also an important nursey ground for the likes of cod and plaice.

Around the UK, however, over ninety percent of our seagrass has been lost in the last century. This is in part due to a wasting disease in the 1930s[34], but its recovery has been hampered by pollution such as sewage and heavy metals. Dredging for scallops, fishing and coastal development have also had a negative impact. Again though, all is not lost. Projects to grow and protect seagrass are ongoing and, with a suitable level of support, could have a noticeable impact both upon biodiversity and Climate Change.

No section on the oceans would be complete without a more significant mention of Phytoplankton. These microscopic marine algae use photosynthesis to produce food and are a key foundation of marine life. They absorb at least as much carbon dioxide as all land plants[35]. They, therefore, play a hugely important role, not just in the oceans but for the whole world. Not only that, but they also influence how much heat the oceans absorb and can even effect cloud formation.

The problem is that the global population of phytoplankton has reduced by about forty percent since 1950[36]. This suggests a rate of around one percent every year which is clearly alarming. It is thought that this is due to rising ocean surface temperatures. Warmer water is less dense that colder water. The warmer the surface water, the less well it will mix with the nutrient-rich colder water below. This means that the phytoplankton do not receive as many nutrients, and so fewer survive.

The best solution here is to both stop emitting so much greenhouse gases into our atmosphere and to remove much of the carbon which is already there. In the meantime, wave powered pumps can be used to mix the water vertically and bring up those nutrient-rich waters. Not only would this increase the number of phytoplankton itself, but also populations across the whole marine ecosystem, all the way up to the top marine predators themselves. It would also increase food security for the millions of people who rely upon the oceans to survive.

This all shows, once again, that we can do it, and we know how. All we need is the political will.

Forests

A forest is essentially a large area containing many different trees and other plants. There are a variety of forest types, depending upon both climate and the array of species.

Forests are often considered to be the lungs of the world. Spend time in a forest and you will be aware of the beauty of nature in all its glory. They are hugely important in terms of biodiversity, supporting an astonishing eighty percent of all land species[37]. These include a tremendous number of different plant species which thrive and survive, supporting all manner of animal life. These include Mountain Gorillas, Jaguars, Capybaras, Green Anacondas, Golden Poison Dart Frogs, Howler Monkeys, Scarlet Macaws and so many more. If you ever get the chance to experience what many consider the quintessential rainforest experience, namely following a Blue Morpho Butterfly along a path, then it will bring joy to your life.

These forests also have over one trillion tonnes of carbon stored within them[38]. Through photosynthesis the plants capture this carbon and release live-giving oxygen. Forests also provide clean water, food, timber, flood protection and ingredients for medicines. They also filter pollution from the air and reduce soil erosion. And there's more; across the world, three hundred million people live in forests, and over one and a half billion people directly depend upon them[39]. There is no doubt that the forests of the world are an awe-inspiring wonder and something to be treasured and protected. So surely we are, right? Let's have a look.

Forests cover about thirty-eight percent of all habitable land mass[40], which is similar in size to that used for farmed livestock. Since the advent of agriculture some ten thousand years ago, we have lost about one-third, or five billion acres of forest. This loss has occurred throughout this time, but half, yes half(!) of this loss has occurred since the start of twentieth century[41]. Not only does this release vast amounts of carbon into the atmosphere, but it reduces the quantity of how much can be absorbed by plants through photosynthesis.

There is also the huge biodiversity loss, with such species as the gorillas, tigers, orangutans and elephants all now endangered. Tigers, for example, number fewer than four thousand individuals, across a few subspecies. This is due to a combination of poaching, habitat loss, conflicts with humans and Climate

Change[42]. The numbers of tigers are now such that it is estimated that there are more captive tigers in the USA alone than in the whole of the wild[43]!

Another critically endangered majestic creature is the African Forest Elephant[44]. This keystone species of the dense tropical forests of West and Central Africa plays a vital role in forest life. Not only do they distribute seeds over long distances, but their dung also provide food for many species. They also create pathways for others to travel along, including for plants to grow. They also create waterholes which others can then use. Again, poaching, habitat loss and conflict with humans are having a grave effect on their numbers.

This is clearly not the situation we had hoped to find. So why are our forests being cut down and burned at such a prolific rate? As we saw in Chapter Four, agriculture, particularly with animals, requires vast areas of land. A great deal of new farmland comes from clearing forests. The cutting down of trees for timber and paper also causes significant losses. There are also the fossil fuel companies who not only destroy forests to extract oil and build pipelines, but whose leaks lead to oil devastating the local environment. Mining for diamonds, gold, copper, and aluminium leads to the removal of even more trees, and can release mercury, which is used in the extraction of gold, into local rivers. This not only kills the fish and other species but affects all those who rely on this water supply, including humans[45]. All in the name of economic progress.

Current rates of loss across the world are well below those in the 1980s[46], but still happening way too much. Whilst it is happening mostly in South America, Asia and Africa, with around half of all losses occurring in the Amazon[47], those of us in other parts of the world cannot point the finger. Firstly, many of the companies causing such destruction are from North America and Europe. These include both those that directly cause the destruction, and those who provide the finance. Secondly, we, through our purchases, are supporting their actions. Clearly, they don't advertise the devastation they are causing so we are generally unaware. It is, therefore, vital that we become educated on such matters so that we can make informed decisions. Lastly, many countries cleared much of their own forest long ago. For example, six-thousand years ago the UK had seventy-five percent of the land covered in forests. This dwindled down to a meagre five percent by the start of the twentieth century[48]. Through conservation and restoration efforts, this has now grown to around twelve percent. Whilst this recent increase is to be applauded, it is still way below the likes of Brazil, which despite huge recent losses, is still at around sixty percent forest[49].

Obviously, the situation is vastly different now to in the past and we understand better the effects of cutting down these vital resources. What I am saying though is that, rather than judging other countries for what is happening there, we need to look at what we can do, both to stop our role within this and support other countries. Whilst this will be looked at further in Chapter Nine, it is worth saying here that most of this needs to come in the form of putting considerable pressure on our own governments as well as the corporations which we purchase from.

We should also add that in some parts of the world the amount of forest is increasing. This can be seen in countries such as China, India, South Korea, Japan, as well as in Europe and North America. This is partly because they have already been through the economic benefits of cutting forests and no longer need to do so. Some of these places also are now relying on resources from other parts of the world.

One example is China[50]. In 1978, China recognised the need to hold back the Gobi Desert. It also noted the effect that deforestation was having both on its soil and weather patterns, with reductions in rainfall on land cleared for agriculture. It has transformed the once, barren, dusty landscape into forest, drawing down carbon and boosting biodiversity. There were issues involved with this process and a need to learn from it. Thankfully, lessons are being learned on the need to mimic nature further with such things as creating a rich diversity of trees rather than more of a monoculture.

In 2007 the Africa Union launched Africa's 'Great Green Wall'[51]. This African initiative aims to build a fifteen-kilometre wide, eight-thousand kilometre long, zone of trees across the entire width of Africa, all the way from Eritrea in the East to Senegal in the South. In combating Climate Change, biodiversity loss, drought, famine, conflict, and migration, it has so far been a clear success. Around fifteen percent has already been completed and has already brought rain, soil fertility, food security, jobs and a reason to stay, to one of the poorest areas on Earth. Planters use a mix of native trees, many of which are drought resistant, to be able to survive in local conditions and support resident wildlife. This amazing story should act as inspiration for us all as we carry out restoration, for a more sustainable planet.

Such afforestation has shown what can be done and will play an important role moving forward, though it is still not as important as stopping deforestation in the first place. Nature itself grows the best habitats but after what we have done, some support, like with Africa's Great Green Wall, can produce positive change. Time is not on our side and whilst in an ideal world we would let nature do its thing, some encouragement along the way can play a critical role. And this can happen all over the world, so what are we waiting for?

Note: afforestation and reforestation are very similar, the difference really being that reforestation is the term used on recently deforested land, whereas afforestation is the term used on land which was deforested further in the past, such as longer than fifty years ago.

To stop deforestation, governments and corporations from across the world must make this their priority. As a vital part of this, support must be provided for poorer countries and communities so that it becomes more profitable to keep the forests rather than cut them down. This could be in the form of payment from richer countries and corporations, including paying locals to protect and reforest as appropriate. Whilst the opportunities for such action are often in lower income countries, we must all play our part and governments from higher-income countries must take their role in supporting this seriously.

Other Habitats

Protecting and regrowing our forests is incredibly important, but nature has provided many other essential habitats too. These include grasslands, wetlands, tundra, deserts and even within soil itself. Each of these support unique ecosystems which play a crucial role for life on our planet. Let's look at each of these in turn.

Grassland is essentially any land in which the grass is the predominant plant, and include the likes of savannahs, prairies, steppes and pampas[52]. These are found naturally in areas which do not receive enough rainfall for full forests, but with more than enough to stop them becoming a desert. They tend to be on open, flat land and exist all over the world.

Well known grassland species include grazing herbivores such as the spectacular giraffes, zebra, antelopes, and bison, through to rabbits and smaller mammals, such as mice. Through this, the grasslands also support their predators, including lions, cheetahs, wolves, and birds of prey such as falcons and owls. Grasslands are not just grass either, they also contain a huge variety of plants, many of them flowering, further supporting many pollinating invertebrates such as bees. Grasslands then, play an extremely significant role in the biodiversity of our planet. They also store significant amounts of carbon, both within the soil[53] and the rich biodiversity which they contain[54].

With all this in mind, our grasslands and the species that live on them must be protected. So, what is happening? Well, sadly, it probably won't come as a surprise to you now that our natural grasslands are under threat. There are many reasons for this including urban expansion, invasive species, hunting and poaching, as well as changes to rainfall and temperatures caused by Climate Change.

The largest current cause of natural grassland loss however is agriculture. Half of all temperate (ones without temperature extremes) and about a sixth of tropical grasslands are now used for agricultural, or industrial, use[55]. But since so much agricultural land is grass, then surely this isn't an issue, right? Well, as mentioned in Chapter Four, in nature, herbivores migrate. This means that they will not overgraze and will spread seeds and nutrients across vast areas. Not so when the land is used for domestic livestock. Also, certain agricultural practices can strip the soil of its nutrients, leaving it infertile for further life. This includes planting monocultures on grasslands and the overuse of chemicals, such as fertilisers and pesticides, impacting biodiversity, both above and within the soil. There are several well-known grassland species which are endangered, or even on the brink of extinction. One example is the Northern White Rhino[56]. There are now just two remaining, both female, and both born in captivity. They were brought together with two males in a sanctuary in Kenya but to no avail, and the last male died in 2018. There are attempts to impregnate the younger rhino using IVF and sperm from a dead male, but this species is virtually extinct. The main causes for this are poaching and habitat loss due to civil war.

Another endangered grassland species is the ground-dwelling Burrowing Owl of North America. In some areas the numbers have decreased by ninety percent[57]. Reasons for this include human encroachment, and the grasslands drying out due to Climate Change. Another major factor is pesticides, which kill insects such as grasshoppers, which the owls eat, becoming poisoned themselves.

Another important feature of grasslands which should be mentioned is that some of them have adapted so well to extreme drought leading to wildfires, that, whilst this may seem paradoxical, they rely upon it to maintain their biodiversity[58]. Fire kills tree saplings and releases nutrients into the soil, which nourishes new growth. It also clears dead plants and leaves so that sunlight can reach the young, emerging plants. Some so-called pyrophytic plants, such as the Australian grass tree, only flower after fire[59], an adaptation which makes use of the newly enriched soil. There is a limit, however. Whilst some fire is good for grasslands in certain areas, too much, such as the increased frequency of fires we have witnessed in the likes of Australia, California and Siberia, is good for no one.

So, what are the solutions to maintaining healthy natural grasslands? Well, all actions to mitigate Climate Change will help. Also, soil erosion and nutrient sapping practices in agriculture need to change, which stems from both education and the right support. Native trees can also be planted to act as windbreaks to reduce soil erosion. Other aspects to reduce soil erosion can be seen in Chapter Four. There can also be controlled burning where appropriate.

Wetlands are regions in which water covers or is present at or near the surface of the soil for much of the year. There are many types of wetlands including marshes, swamps, lakes, rivers, deltas, and peatlands. These contain a huge variety of biodiversity. Wetland species include hippos, flamingos, capybara, and beavers. They are also significant sinks of carbon. Whilst they cover just six percent of the Earth's land, they store over ten percent of the terrestrial carbon[60].

Wetlands also act as a filter for pollutants, such as nitrogen from fertilisers. If wetlands are drained or reduced, or even if the fertiliser runoff from farms is too much, they can no longer provide this filtering action, leading to large amounts seeping, leaking, and running into the oceans. This can lead to eutrophication on a huge scale, robbing all dissolved oxygen from the water, creating literal dead zones, such as within the Gulf of Mexico[61]. Again, their importance is not in dispute, so what is happening with them?

Well, worldwide, around sixty percent of wetlands have been destroyed in the last hundred years[60]. This includes up to ninety percent of those in Europe. These are often drained to provide land for agriculture and housing. They can also be seen unfavourably as a breeding ground for mosquitos. This draining both releases carbon into the atmosphere and has a negative impact on the rich wetland biodiversity.

One endangered wetland species is the European Water Vole. Living wherever water is found, such as streams, ditches, marshes, and lakes, in the UK the Water Vole had decreased in numbers by a massive ninety percent by the 1990s[62]. This decline has continued despite conservation efforts. Whilst sometimes confused with rats, they are nothing of the kind and play an important role in the local ecosystem. For example, their burrowing encourages microbial activity in the soil, providing nitrogen for plant growth. Reasons for their numbers plummeting include habitat loss leading to population fragmentation as the populations become isolated. This causes lower genetic variation within each population. Waterway pollution from agriculture is also damaging their habitats, and thus the water voles themselves. Another major factor is the introduction of an invasive species, namely the American Mink, to the UK in the 1920s. These were brought over for fur farms (I know!) but many escaped. They are now widespread across Britain, and whilst there is some indication that mink numbers are decreasing, the fact that they can fit into water vole burrows and wipe out entire populations, has had grave consequences for the beloved water vole, encapsulated as 'Ratty' in The Wind and the Willows.

Since we are discussing wetlands, it seems only right to mention the fortunes of the European Beaver. This native species was once widespread across Britain but became extinct in the sixteenth century due to hunting, both for their fur, and for castoreum, the liquid which greases their fur, and is mixed with urine to mark their territory. This substance is used in perfumes, food, and medicines. In 2009 a trial ran in Scotland to reintroduce the beaver back into the UK[63]. This was successful and there are now two populations in Scotland and another in England. They have also been reintroduced at hundreds of other locations across Europe. As an important part of the ecosystem, beavers naturally coppice, allowing sunlight for plant growth and homes for insects and birds. Their dams also support the habitats of many other species, including otters and water voles, as well as creating breeding grounds for fish and insects, leading to greater numbers of insect eating birds such as spotted flycatchers. Not only that, but the dams also regulate the water flow, thus helping in times of drought. The dams can also capture some of the runoff from agriculture, leading to an improved water quality down-stream. Clearly this reintroduction has been a real success and, once again, shows positive changes that we can make.

If all this wasn't reason enough to support wetlands, then consider the fact that they also bring not only recreational opportunities but provide significant flood protection. This is something that's always important but will become even more so as Climate Change increases around the world. The wetlands also filter pollution[64], acting as a natural wastewater treatment plant. Amazing! Imagine if the water companies, as discussed in Chapter Five, also stopped dumping raw sewage into the waterways.

Whilst we won't be discussing each type of wetland, it would be remiss not to mention peatlands. Peat is a thick, wet fossil fuel, that has developed over hundreds or thousands of years. It contains a vast amount of dead and decaying matter as well as supporting living organisms. It encompasses so much carbon that is has double that of all the forests in the world[65]!

Peatlands contain so much carbon due to the plant material they contain. Within the peat, dead plants decay without oxygen. Unlike landfills, however, most of the methane cannot get out and the overall effect is that peat absorbs far more carbon than it emits. The issues arise when the peatland is drained of water for such human activities as agriculture, housing or removing peat to be used as fuel or soil. When this occurs a great deal of the carbon combines with oxygen in the air, becoming carbon dioxide and being emitted. The previously produced methane is released too. Thankfully there are calls to ban the removal of peat, with peat sales being banned in the UK, for example, from 2024.

Another important habitat which should be discussed is tundra. Tundra are large, treeless, cold regions, with little precipitation, long winters, and short summers. They are characterised by permafrost, in other words soil which remains frozen for many years.

Despite the lack of trees, tundra do support low-to-ground vegetation, such as grasses, mosses, lichens and shrubs[66]. These support a variety of fauna such as Lemmings, Arctic Hare, and Musk Ox. So, whilst the biodiversity isn't as large as in some other regions of the world, they do support species not found anywhere else, and thus, are an important habitat.

The permafrost also stores huge amounts of carbon, estimated to be around twice as much as currently in the atmosphere[67], with some of this carbon having been stored for at least six hundred and fifty thousand years[68]. As this permafrost melts, microbial activity increases, causing the organic matter stored within it to decompose. If oxygen is present, then this releases carbon dioxide, which is bad enough, but in much of this soil, oxygen is not available, meaning the decomposition will lead to the release of methane. Even now, summer methane emissions from tundra are ten times their normal rate[69]. As the Earth continues to increase in temperature, this will further accelerate, leading to even greater warming.

Now here comes a, perhaps surprising, link between biodiversity and the melting permafrost. Like all regions around the world, the tundra experience higher temperatures in the summer, causing the permafrost to absorb heat. When the snow later falls on top of this soil, it acts as a layer of insulation, thus keeping the heat in and causing greater decomposition. When this then thaws, more methane is released. Ideally then, there wouldn't be thick layers of snow on top of the surface. This is where the biodiversity comes in.

The more grass, rather than trees, on the surface, the more food for herbivores such as the Yakutian Horse and the Musk Ox. The large migrating numbers of these species break up the surface snow, leading to colder permafrost and fewer emissions[70]! Whilst vast herds used to roam this land, they were hunted, hugely reducing their numbers. Efforts are now occurring to restore them to their natural habitat. This is not only a solution to increase biodiversity in the region, but also to mitigate Climate Change. An amazing example of how different solutions fit different parts of our planet.

One more habitat to discuss in deserts. Deserts are relatively barren landscapes, characterised by little precipitation. Hot deserts, which most of us are more familiar with, include examples such as the Sahara and the Gobi deserts. There are also cold deserts, with both Antarctica and the Arctic being larger deserts than the Sahara[71].

Whilst having a lower biodiversity than other regions of the world, deserts still host a wealth of plant and animal life, including Sand Cats, Jerboas, Snow Leopards, Meerkats, and the unbelievably cute Desert Hedgehog[72]. Not only do they have their own inherent beauty and importance, but many of their species are also important sources of food and medicine for local communities. So, having some desert is important, but what we clearly don't want is desertification.

Desertification is the changing of land so that it can no longer support as much plant growth as it could in the past, such that it is not easily reversed. In other words, its where fertile soil becomes so infertile that it will take a long time to recover[73]. This is caused by deforestation, over-farming, drought, fire, housing, and mining. The subsequent soil erosion not only removes life-giving nutrients, it also reduces the ability of the land to retain water. Whilst this has always occurred, it is now estimated that desertification is taking place at over thirty times the historical rate[73]!

Around two billion people around the world live on drylands that are vulnerable to desertification[74] and, like many of the world's problems, it has the greatest impact upon the poorest and most vulnerable. As the land turns to desert, it no longer supports these populations, leading to the displacement of people, just so they have a chance to live. With warming global temperatures, and increased drought in many, already very dry, places, this will intensify and lead to the mass migration of people.

This is why projects such as the aforementioned African Great Green Wall, and others such as the food forest on the Greek island of Evia[75], are so important. They not only increase biodiversity but breathe life back into the soil itself. This makes a hugely positive contribution to local communities. Other initiatives, such as stone barriers to reduce water runoff are being implemented, yet desertification continues. We are all part of the cause and so must all be part of the solution.

From Pollinators to Pangolins

Pollinators carry pollen from one flower to another, thus allowing for reproduction. Whilst some plants pollinate through wind or water, ninety percent of all flowering plants rely upon animals, such as bees, wasps, butterflies, moths, birds, and bats for pollination to occur[76]. Without these pollinators, many plants would soon die out resulting in the collapse of entire ecosystems. Indeed, it may well be that plant-pollinator relationships are some of the most important on the planet. Not only is this vital for biodiversity across the Earth, but huge numbers of crops, including many fruits, vegetables and nuts, also rely upon such pollination, so if anything happened to the pollinators it would directly threaten our own food security. So, what is happening?

It breaks my heart to report that pollinator numbers are decreasing around the world[77]. The main reasons for these are habitat loss and agriculture, with the like of monocultures and pesticides playing a significant role. Even when pesticides, which are shown to be kill bees, are banned in one country, they are still used in others. One example is neonicotinoid thiamethoxam which was banned in the EU in 2018, but can still be used, under certain circumstances, in the UK[78].

But there is more. Climate Change is now affecting the times in the year when pollinators are active and when flowers are in bloom[79]. One example is the UK's Early Spider Orchid, which relies upon Miner Bees for pollination. Their needs are so in-tune that they have evolved so that the bees emerge at the same time as the orchid flowers. Both are triggered by environmental factors, and rising spring temperatures mean that, whilst both are now out earlier in the year, the timing of the bees has been affected to a greater extent than the orchids. The upshot of this is that they are no longer in sync. This obviously reduces pollination of the flowers and food for the bees, damaging the populations of each.

Similar mismatches have been observed elsewhere, such as between Oak Tree buds and Moths, and between Red Admiral butterflies and Stinging Nettles. Other mismatches are also suspected, such as certain sea birds and the fish they prey upon.

What should be clear is that human actions are impacting upon pollination. The pollinators of the world have such importance ecologically, economically and in supporting our needs, that we must do all that we can do to support them. More on this in Chapter Nine.

Finally, let's look at the plight of the Pangolin. These 'scaly anteaters' of Africa and Asia are vital parts of the ecosystem, consuming great numbers of insects, particularly termites, which would otherwise explode in number. But why have choose the Pangolin to discuss here? Well, it's because more Pangolins are trafficked illegally than any other mammal[80]. Not only are they hunted for meat, but for the supposed medicinal properties of their scales. Despite increased restrictions on international trade, numbers are such that the eight different species of pangolins range from vulnerable to critically endangered. How is hunting a species to near extinction logical?

But there is more. It has been proposed, but not proven, that Covid-19 passed from Pangolins to other species, and then onto humans[81]. Whether this ever proves to be true or not, what is clear is that the pangolin is not to blame. They never chose to be trafficked and exploited, we did that, and we are all part of that system. We have already looked briefly at

Covid-19 in Chapter Four and will do so more in Chapter Eight, but for now let's stop blaming the animals.

Summary

Our natural world is glorious in its diversity. It brings awe and beauty into our lives, and the vast range of species across the planet support our very existence. Do we really want a world where our children grow up unable to ever see such majestic creatures such as Tigers, Elephants and Polar Bears? How are you going to look your children in the eyes and tell them why that opportunity was stolen from them? Not only that, but the greater the diversity the greater the chance for life to not only survive but thrive.

When you look at the suffering of others, whether humans, or other species, how does it make you feel? The reason most of us feel distressed when others suffer is because, at a deep level, we know that we are not separate from them. That pain tells us who we really are. It tells us something positive about ourselves and our connection to nature. It should give us hope that we can do what is right.

We have seen that human actions have caused devastation to so many species on Earth, but it is not too late. Whilst we will look more into solutions in Chapter Nine, consider what you are willing to do, and indeed, not do, to support a better world for all. Remember, this is our climate, our future, and there is a better way.

CHAPTER 8

PEOPLE POWER

So far, we have seen that the current environmental issues are very real and extremely serious. We have also seen that we know the solutions, but they are not being implemented anything like as much as they need to be. But why not? In this chapter, we look at the role which humans across the world have had on all of this, who is affected most, how different people are reacting to the crisis, and what difference it is all having.

It is important to recognise that some humans have always divided people into distinct groups, not only to obtain power over one another, but also to give themselves more identity[1]. These groups include gender, race (a social rather than biological construct anyway), sexuality, class, nationality, religion, or age, amongst many others. This division of people into different factions can lead to some groups seeing themselves as superior to others, causing discrimination and mistreatment of others, even leading to segregation and war.

Whilst this is still occurring, our Environmental Crisis provides us with an opportunity. It's too big. None of us can resolve this by ourselves. As the evidence becomes ever more overwhelming, it will become clearer that it is only through working together that we can do enough to significantly mitigate Climate Change and our other environmental issues.

We are one people living on one planet. No, more than that, we are one vast group of different animals, plants, fungi, and other kingdoms all sharing this world. Of course, there are differences but that is part of what makes the Earth such a wonderful place; it is these differences that bring beauty into our lives and make the whole world stronger. The greater the diversity, the greater the chance of survival. This is a natural law and should be celebrated. So, let's have a look the forces at play which have stopped us from coming together, and what we can do about it.

Are People Essentially Good?

If most people, in their heart of hearts, are good people, genuinely care for others, and truly want what is best for others, whether their loved ones, people across the globe, and indeed for future generations, then there is hope indeed for humankind.

If the opposite is true, if most people aren't inherently good, if we only look out for our own, immediate, self-interests, then we won't do anything until we are under immediate threat ourselves. By which time it will be too late. So, let's have a look.

Studies have shown that, thankfully, most humans are inherently good[2], and have a desire to help each other out. This heart-warming realisation should do much to lift us up. Most people want to do the right thing. So, let's now investigate this further, to understand the role of people in all of this and how we can further hope in our hearts.

What we know is that most people know at least a little bit about Climate Change, plastic pollution, and other environmental issues. It is also true, however, that many don't realise just how serious it is. And can we really expect them to? After are hard day at work or school, it's far preferable to spend nice time with our loved ones, with family and friends, watching tv, doing sports, or many other activities. Why would most people want to spend their time learning about and contemplating our Environmental Crisis? It's hard. None of us want it in our lives. Much better that it doesn't exist, and if that's not possible, then maybe, just maybe, we can hope that others will sort it, so we don't need to worry. No, blaming others for a seeming lack of engagement helps nobody. We have not lived their lives, we are not living their life, so how can we judge them?

On top of this, governments and media so often tell us, whether through their words or their actions, that everything is in hand, that there is no need to worry. Much nicer to go along with this, rather than consider the possibility that those we have put faith in aren't looking out for our best interests. That really would be hard on us.

We should just note that people who are narcissistic are more likely to either be politicians or take a more active role in politics[3]. In other words, people who have an increased sense of self-importance, a sense of entitlement, a lack of empathy, a high level of arrogance, a need to be admired and an inability to take criticism, and find it difficult to see others' perspectives, are more likely to seek positions within politics or other positions or power. This suggests that whilst only around one in two hundred people are narcissists, the percentage involved in politics is higher. Whilst this clearly doesn't mean that all politicians are narcissists, it is certainly something to keep in mind.

History

As we saw in Chapter One, Climate Change is real, happening right now, and caused by human actions, mostly the burning of fossil fuels. We also know that we are at a point in history in which significant action is needed, both to reduce and reverse further warming. We must also adapt to what is already occurring and is likely to continue. Some of these actions may seem quite drastic. Of course, it would have been so much easier to have started earlier and at a more gradual rate, but maybe we didn't know? So, here we should look at how long we have known of these issues.

Well, by the eighteenth century it was known that humans could affect climate, through activities such as the clearing of forests. For example, it was recognised that the climate of North America was undergoing rapid change due to settlement by Europeans, and the changes occurring would make it suit European 'civilisation' over 'primitive' native cultures[4]. Even the use of words civilisation and primitive tells us plenty about the attitudes towards other cultures at the time.

The eighteenth century also saw the beginning of the industrial revolution, initially in Britain, with a move to an economy dominated by machine and industry[5]. Before this, most people lived in the countryside, and made things in their own homes. The large machines which arrived during the industrial revolution, however, could not be powered from the home, as manufacturing moved to factories[6].

In the early eighteenth century, Abraham Darby discovered that coke, which comes from coal, could be used to smelt iron. Before this time only very small amounts of coal were used. A move then came to power the machines in factories with coal. This meant that the factories no longer needed to be built along rivers to obtain the power of moving water. It also meant that fewer trees would need to be cut down. Some clear benefits here. This, along with the invention of steam engines, and coal powered lighting saw an explosion in the use of coal[7]. But was anything known about its effect on the environment?

Well, in 1842, Samuel Forry analysed data from the US army, which suggested that Climate Change was not occurring[8]. The previous ideas of human changing the climate were then rejected. Then in the 1850s, both Eunice Foote[9] and John Tyndall[10] investigated the properties of a variety of gases within the atmosphere. They, independently, noted that whilst oxygen and nitrogen let infrared radiation pass straight through them, others, such as carbon dioxide and water vapour, were very good at absorbing the energy from the infrared. This was the first recognition of the Greenhouse Effect and the realisation of the importance of the composition of the atmosphere's gas in regulating the Earth's temperature.

Svante Arrhenius then took this further, and in 1896, determined that relatively small increases in carbon dioxide within the atmosphere, would have a significant impact upon the Earth's temperature[11]. He conducted experiments and carried out calculations, which suggested that a tripling of carbon dioxide could increase Arctic temperatures by at least eight degrees Celsius.

Let's just take a moment to take in this information. Whilst experiments have been improved, and calculations moderated, it was known one hundred and thirty years ago that burning fossil fuels, such as coal, directly leads to an increase in temperatures on Earth!

Later, in the 1930s, Guy Stewart Callendar investigated evidence from weather stations across the world and determined that there had, indeed, been a measurable increase in the Earth's global temperature[12]. So, this was known over ninety years ago!

Over the years, even more evidence came to light, but many people still dismissed any concerns, since it was believed that carbon dioxide released from burning fossil fuels would be absorbed by the oceans. Then in the late 1950s, Charles David Keeling, showed that the concentration of carbon dioxide was indeed increasing[13]. Keeling then pursued this work over the years, leading to the so-called 'Keeling curves', which clearly show the increase in atmospheric carbon dioxide.

When this was matched with measurements in rising temperatures, the evidence was clear; Human activity was, and still is, causing the increase in carbon dioxide in our atmosphere, the effect of carbon dioxide on global temperatures in well understood, and this corresponds to the increasing global temperatures that are being observed.

And that's not all. When we include further evidence, such as cores in sea-beds and ice, past levels of carbon dioxide, evidence from other planets such as Venus, and intricate computer models, the evidence for anthropogenic Climate Change is overwhelming. Faced with such evidence, it really is a struggle to understand, without the kind of filters and misinformation that we saw in Chapter Two, how anyone could still deny the existence of human-caused Climate Change.

Communicating the Science

We understand Climate Change and the other environmental issues by looking at the evidence and developing theories and models which we can test, through experimentation, and improve. Through this process, our knowledge and understanding of the situation is enhanced. In other words, it is based upon science.

The problem as to why more people are not on board with what is happening is nothing to do with the science itself. That is sound. Indeed, there is now more data and evidence on Climate Change than almost any other scientific pursuit. So, what is it to do with then? Well, in Chapter Two we saw how we all react differently to overwhelming information. In this section then, let's look at how the science is presented, and indeed interpreted.

Firstly, we should recognise that none of us are passive receivers of information. How each individual's mind works, along with their life experiences, and the influences of others, leads to the receiver applying meaning to this information. We can all be given the same information, but interpret it in different ways.

This essentially means that just stating the facts can never be enough. As we subconsciously apply our own mind's filters, then we, ourselves, distort and revise the information, into a form that makes more sense to us and fits better with our own world view. Therefore, as mentioned in Chapter Two, we should never get angry with, or blame, those who deny, as frustrating as it may sometimes be. Let's have no more 'us' and 'them'. Even with those who deliberately spread misinformation, let us strive to work with, rather than against them. The 'us' and 'them' view just forces the choosing of sides, from which can be very difficult to turn against.

So, if explaining the science, stating the facts, sharing the evidence, it not enough, then what is? Well, there is no doubt that there is a rise in anti-science in many parts of the world[14]. In other words, an increase in the number of people rejecting mainstream scientific views and replacing them with theories that can often be seemingly based on (a small amount of) evidence, but when investigated further, are far from the truth. Examples of this include not only the rejection of the overwhelming evidence on Climate Change, but also evolution, Covid-19 vaccines, and even believing that the world is flat!

This isn't helped by the fact that some people look down upon those who reject the science, believing themselves to be 'better', 'more intelligent' or at least 'better educated' than those who deny. But there are also plenty of examples of highly intelligent, highly educated individuals, who, for one reason or another, have filters that reject the evidence from climate scientists[15]. Treating science-deniers with disdain might make you feel better about yourself in that moment but can never bring about understanding. As Yamamoto Tsunemoto said, "By bringing shame to a person, how can you expect to make him a better person?"[16]

Like many things in life, it is all about probability. Having a scientifically literate population increases the likelihood that people will look at the science themselves and understand it, rather than reject it[17]. This links with the section on education later. It is important to note at this point, however, that being scientifically literate isn't about making people think the same thing; it is about making people more aware and able to delve past the bold type to the actual evidence, so that each person can make their own evidence-based decisions.

Another important factor in sharing scientific information is not just its accuracy, but also hitting the right level so that it is neither over-simplified, nor over-complicated. As a teacher I see this every day. If too easy or too hard then the students will turn away from it. This is further complicated by the fact that the correct level is different for all of us. In a school, a skilled teacher can use differentiation to further the understanding of all, but this is far trickier when presenting information to the public. It is also easier to find fault, real or not, in information which is too simple or complicated. I am confident that I haven't got the levels quite right for many of you reading this book, but this just further highlights the importance of accuracy and level of engagement, discussion, and investigation, rather than the mere transfer of information from one to another.

Two significant sources which affect our attitude and understanding are media, including social media, and our education. Let us then look at these two in turn to recognise their role and how their manipulation is affecting us all.

Media

When we say media, we are talking about the variety of ways in which information is communicated to us, be that through tv, radio, newspapers, billboards, and of course, the internet and social media.

Before we look further into the role of the media, it is important to understand who owns, and ultimately controls it, in its various guises. Essentially the answer is either the state, or the private sector.

With state-owned media, such as the BBC, in theory it is independent of the government, but the Chairperson and several board members are appointed by the Queen, on recommendation from the government at the time[18]. The government also can apply pressure regarding their funding, even suggesting the loss of finances for not doing as the government would wish them to[19]. State owned media in other countries can be far less subtle.

When in private hands, the main newspapers, for example, are owned by extremely wealthy individuals who have their own political agendas. They generally push a narrative to convince people to think a certain way, and to vote a certain way, such that fits best with their status and position. For example, one extremely wealthy family own three of the biggest newspapers in the UK, some of the largest TV networks, newspapers, book publishers and magazines in the USA, including Fox News, as well as an even greater range of TV networks, newspapers, and magazines across Australia[20]. And the message from each of them, to varying degrees, is right-wing in nature, dismissive of Climate Change, skeptical on vaccines, and highly influential on how millions of people think, act and vote.

Well, what about social media? This is something that we can all contribute to, so surely, it's ok, right? Well, the number of different social media outlets that most of us use is pretty small. And their principal goal is the same as all large corporations, to make profit. So how do they do that? Well, they make money by selling advertising space. The thing is, no company is going to pay for something that no one is going to see. So, the social media companies are designed to make it more likely that you spend ever more time on their platform, and they can target adverts that you are more likely to interact with. When you comment, share, click, or otherwise interact on that social media platform, all this information goes into an algorithm which gives you even more of the type of information you have already shown an interest in[21]. Search engines fit into this too. Based upon what you have searched for previously, different people can type in identical first words and receive completely different suggestions for the rest of the sentence.

A consequence of this is that it essentially creates an echo chamber, so that whatever world views you currently hold, however strongly or weakly, are fed again and again, reiterating similar messages to reinforce that view, further separating us from each other. If pretty much all we see is so one sided, then everyone else must be wrong, right?

Added to this, evidence suggests that media has the greatest influence over people's thinking when society is deemed to be more vulnerable, whereas it is less influential in times of social stability[22]. Take what you want from this, but it does tell us that if the media want a larger influence over what we think, then it is in their interest to have an insecure and unsteady population.

It is also important to recognise that it is the media which controls the vast majority of what we learn. Scientists, for example, are only partially in control of the way in which science is communicated. In other words, the science appears in the media, on the media's terms, not the scientists'. Science, along with all other issues, is subject to 'framing'. In other words, to a way of delivering the message in such a way as to affect how the receiver interprets it[23]. Let's now then look at the ways in which science is framed, or presented to the public[24];

The 'Scientific' frame is just stating the science, which we have already seen to not work for many people. It also can be written in such a way as to undermine the issue, with some, often deliberate, cherry-picking leading to a ridicule of the science. Note that when this happens, what they are essentially stating is that one journalist understands the climate far better than pretty much all climate scientists, but of course, it is never stated in such a way. This framing also separates the science from politics, social and cultural issues, putting into the subconscious that science should not have any influence over such things. There is also the issue that sometimes things which are not true are stated as fact, or that there is a deliberate omission of the truth. This, as we saw in Chapter Two, can sow doubt in people's minds. It can even lead to a mix-up of science and technology, suggesting our wonderful scientists have worked out the issue but they will now solve it all with technology alone; someone we noted in Chapter Six to be incorrect. Furthermore, the inherent uncertainty within all science, can be oversold, suggesting that scientists don't really know if Climate Change is real or not. Whilst we saw in Chapter Two that this is not the case, it can help justify inaction on climate to those who don't want, or are fearful of, change, and that their world view is still ok.

There is then the 'National Security' frame. Here we see concern for just our nation. For example, when looking at the move away from fossil fuels, this may be framed as a way to become less reliant upon others, such as oil from the Middle East. It can also cover articles on increasing migration (they normally say 'migration' rather than 'climate refugees') which leads some people to feel we should do more to keep our borders 'safe'. Within this the emphasis on 'them' and 'us' rather than one body of humans is clear to see.

Then there is the 'Polar bear' frame. This is predominantly aimed at wildlife lovers and includes some of the wildlife programmes that so many of us enjoy. Often, they show some of the effects of Climate Change on wildlife, without actually going into the real causes and the big scale solutions that will make a significant difference. These programmes also rarely include the impact upon people, and how their lives and being affected by Climate Change and our other environmental issues right now.

There is also the 'Money' frame. This includes information about how much the proposed solutions to our environmental issues will cost us and how we can't afford them. Far less is shown on how much it will cost if we do nothing, or not enough. Also, here we see suggestions that Climate Change and the like are not based upon science but are simply tools of anti-capitalists who wish to overthrow capitalism. This frame is also often used by politicians and the private sector to justify their actions.

The 'Catastrophe' frame is all about the dire outcomes of our actions, or lack thereof. It can certainly sell papers and make people sit up and take notice, but it can also be exaggerated to the extent of doomism, where people are so overwhelmed by the information that they see no point in making positive change, as they may see it as "too little, too late".

The 'Justice and Equality' frame looks at who has done what and who has the most responsibility for changing their behaviour for the better. Whilst it can be useful, it can also lead to the 'blame game', and often statistics are used by those wanting to blame others rather than take responsibility themselves. It can, however, act as a call to action for those with high levels of empathy and strong ethics.

Lastly, the 'Celebrity' frame is all about putting the attention on those with a high profile, such as Greta Thunberg or David Attenborough. This puts more faith in that one person rather than the evidence itself. This frame can be used to either galvanise people, or to demonise or ridicule those individuals, as has been done to Greta on many occasions, in an attempt to undermine the whole issue.

We should also not underestimate the impact of TV shows, movies and music on our perceptions. Hollywood, for example, loves an apocalyptic story, and some, such as The Day After Tomorrow, suggests that highly extreme events were already predicted to take place, which isn't true. This creates the suggestion in our subconscious that Climate Change is not so real after all.

Throughout all of this we should consider whether media can ever be truly objective. And the answer, of course, is no. None of us are blank slates. We all come with our own baggage of experiences, thoughts, and ideas about the world. Even when striving to just state the facts, they are selected and framed in such a way as to present a particular viewpoint, however conscious or not we are of this happening. This very much includes me here. The aim should be, surely, to find and speak the truth, and base it upon evidence, no matter how uncomfortable that may be. Ideally, the evidence would look at both details and the big picture, not just hand-picked aspects which further our own world view. It is a challenge indeed, but there are fewer which are more important.

Whilst even the most well-meaning presenter of information has their own built-in bias, this is very different to the deliberate and calculated spreading of misinformation. The willful transmission of inaccurate information carried out by the fossil fuel industry, along with aspects of the media, is well documented within certain circles[25], and yet is something that is rarely presented in mainstream media. In Chapter Two we looked at the dangers and reasons behind such tactics so there is no need to go over them again here, except to reiterate that this is extremely well funded by powerful people who have huge influence. Now consider, are they doing this to benefit you, or themselves?

These days the evidence is so overwhelming that most people understand Climate Change is real and caused by human activities. As people become ever more aware of the issues, we may turn away from certain activities, products and political parties. One tactic now deployed to combat this is known as greenwashing. This is 'behaviour or activities that make people believe that a government or corporation is doing more to protect the environment than it really is'[26].

Greenwashing varies greatly in terms of the strategies used, from 'subtle' labelling on packaging to entire campaigns which suggest that the corporation is a protector of the environment. The examples are far too numerous to go through here, but a quick search can easily find some of those which have become well known publicly. These include Volkswagen deliberately fitting a device to trick tests, when in reality the emissions were far higher than those permitted, and affected eleven million cars[27]; fossil fuel companies, such as BP, Exxonmobil and Shell, making commitments such as reducing emissions that, when you look at the details, do not actually include the emissions from their products![28]; Coca-Cola promoting itself as an environmentally friendly company whilst continuing to be the largest plastic polluter on the planet[29]; and financial institutions, such as banks like JP Morgan, and Barclays, portray sustainability whilst still investing heavily in fossil fuels[30]. And what is the media saying about this? Well, not a lot. You can find the information, but you have to really look for it and much of the media makes no mention of this at all. This means that most of us, through no fault of our own, remain unaware.

This isn't helped, of course, by both government support for such corporations and the vast resources used to employ law companies to protect them[31]. The people who run these organisations know that they are damaging the environment, but profits and shareholders much come first. Any loophole in the law, however unethical, must be explored to put money and power first. It's not that they have a malicious intent to harm our world, it's just that it cannot get in the way of their profits!

And what is the media doing to inform us about all of this? Well, as mentioned, since the big media is run by those who wish to maintain the status quo, then very little. Some are more explicit that others. Others, such as the BBC, portray themselves as 'balanced' and may interview both a climate scientist on one hand, and a climate denier, often supported by the fossil fuel industry, on the other[32]. Both are presented as having equally valid views, despite one being based on an overwhelming wealth of evidence, and the other, quite frankly, is not. It is not 'balanced' to have one expert share peer-reviewed scientific findings and someone else state that it's not true; it is a deliberate policy to create doubt.

Having said all of this, there have been articles concerned about the climate, from as early as the Rodney & Otamatea Times in 1912[34]. Just let that sink in a minute. Well over one hundred years ago there was an article warning about the effects of burning coal on the temperature of the Earth. And yet so little has been done about it. What is lacking is the political will. Let us now then look at why this is so. We know that the mass media is controlled by extremely powerful governments and private individuals who wish to play down the true significance of Climate Change, but what about the political and economic systems themselves?

Governments and Systems

It is said that a government acts to protect the people. If this is true, alongside the overwhelming evidence that we have seen on Climate Change and the host of other environmental issues, then surely acting upon this is a major priority of governments across the world. Well, since they have been going on, and we have known about them, for such a long time, then clearly, they aren't doing enough. But why not?

Remember that the government is made up of politicians. These politicians are people who have become, or already were, public figures, with a great deal of influence. But how does someone become well known and elected, even at a local, let alone a national level? Well, to have a chance of election, a good deal of campaigning needs to take place. A successful campaign requires a great deal of time, energy, and money. Most individuals work hard to earn their money so they can pay their own bills, so won't be able to also run a successful campaign. The only real chance is to have significant support from others, whether they be individuals or organisations. Generally, those who have enough finances to be able to support such a campaign are those who are already benefitting greatly from the current system. These financial supporters of politicians and their campaigns, include industries such as fossil fuels, animal agriculture, banking and even arms manufacturers. Why would they then support a candidate that intends to bring about real change? When you add in the influence of the media on this, who we have seen, are also run by those wishing to maintain the status quo, then it is no wonder that the vast majority of our elected representatives are those who have no desire to bring about the change required to make a genuine difference.

And what about once they are elected? Well, as has been mentioned previously, political parties receive significant amounts of money from powerful bodies such as the fossil fuel industry[35], the food and drink industry[36], and the financial sector[37]. Again, supported by those who have vast money and influence within the current economic and political model.

Many of us around the world may well be living in a so-called democracy, where people vote to decide who is in power, but let's not be deluded. As long as we have a system in which political parties and individuals require support from powerful organisations in order to achieve success, and that the media is also controlled by those similarly powerful, then how can we ever expect real change? We should also note that modern democracies are based upon competition. The goal of a politician is to act in such a way as to defeat their opponent(s) and get elected. It is based upon division, further enhancing the idea of them and us; the very opposite of what we need when dealing with the big issues.

The whole mechanism is so well-oiled (pun intended) that many of us don't even recognise that our actual choice when we vote is so limited. The system in most countries across the world works in this way. No wonder protestors call for 'System Change, Not Climate Change'.

Now don't get me wrong, like energy resources, some political parties are way, way worse than others, and so swapping one for another can bring some positive change, it's just that whilst we have a system in which big business and the media have so much influence over who is genuinely up for election, and how we think about them, then real change will be extremely difficult to come by. In many countries around the world there are just two or three parties with enough support from the business world and media, such that they are in the public eye enough and perceived to be 'not a wasted vote'. No one wants to think that they have no impact on the world, and so most people will vote for those who 'have a chance of being elected'. We can see how the issue is self-perpetuating.

Within this however, every now and again, a candidate appears who seems to have a deep passion to bring about actual change. Change to the system. This genuinely terrifies those in power and so this candidate is vilified, at every turn. They are made to seem like the bad guys that will bring ruin upon us all. And how can most of us not go along with this? We don't have the time, energy, or resources to investigate the truth for ourselves, so we go along with what our chosen media is telling us.

This section of this chapter could be a full book in itself, and I don't have all the answers. But what is clear is that if the current system cannot produce the real, positive change required, then it is the system that needs to change. Until then any changes will be minor and continue to benefit those in power more than the rest of us.

Ok, so the political system needs to change, but what about the economic? Well, the prevailing economic model incorporated across much of the world is capitalism. That is, one in which property, business, and industry are controlled by private owners rather than by the state, with the purpose of making a profit[38]. True capitalism would have free markets, but given the huge subsidies given to certain industries, then the markets are anything but free. If they were, then the prices of the likes of fossil fuels and animal agriculture would be much higher, meaning we would use them less. Then there is the responsibility that companies have to their shareholders, further driving growth and profits. Through this it should be very clear that the current model requires corporations to continue to make a profit[39], meaning that ever more goods and services are required.

Built into this is a constant bombardment, from birth, about how our lives will be better if we just had a bigger house, a faster car, the latest phone, etc. Materialism, required for those who run corporations to make vast profits, is so commonplace that many of us hardly notice. And it's not just that this strive for more and more is damaging our world, studies have shown that increasing materialism in children leads to more unhappiness and dissatisfaction with life[40]. Surely this is something that none of us want.

We also saw in Chapter Five that we are already surpassing the sustainable limit of our planet ever earlier each year. How can we then hope to foster a true model of sustainability within a system that requires perpetual growth? Well, clearly, the answer is that we can't. This may well not be what we want to hear, but that doesn't stop it being the truth. The science simply doesn't allow this to occur. How could we ever have expected infinite growth on a finite planet? The Earth's resources are limited, and we must find a way to work within these, rather than continue to push ever further past them. How can we justify not only destroying so much for future generations, but also for people around the world right now?

Honestly, if someone wanted to design political and economic models with the intent of destroying much life on Earth, making some people ridiculously wealthy and powerful, and so many others in poverty, all whilst ensuring the majority don't even realise just how bad it is and therefore continue to support that very model, then they couldn't have chosen much better than the ones we have right now.

At this stage some people will criticise these words for being anti-capitalist, but this misses the point entirely.

We literally cannot afford, as a species, to continue to plunder the Earth's resources at anything like the rate we are currently doing, let alone increase it. This is just basic science. Whichever system we use, our use of the Earth's resources must reduce. The Earth can become sustainable once more, but changes need to occur for this to happen.

Clearly then, both our current political and economic systems need to change. So, why isn't this happening? Well, those who currently gain the most not only don't want it to change but use the media to convince us that it doesn't need to. But it's more than that, the reporting is done in such a way that many of us don't even realise that it's the system that is the issue. Instead, we are just told that we as individuals are to blame, and it is us who need to do more. Or even, as a perceived protest, there is anger directed towards the current government, as if a simple use of 'democracy' to vote in a different major political party, again supported by those who want to maintain the status quo, will bring about the real change that we don't even identify to be required! Don't get me wrong, many governments are complicit within this and so need to change, but just switching from one major political party to another will rarely be enough.

It is also stated that the measures required for significant climate action would be too expensive and damage the economy. Imagine telling that to those who are already dying, losing loved ones, their homes, their livelihoods, their very dignity. If it is our economy that we are most concerned about then maybe it's our economic system that's the issue. Let's then have a look at how expensive it would be compared to not doing anything.

In 2006, commissioned by the British government and led by economist Nicholas Stern, the Stern Review[41] was carried out. This extensive review looked at a wide range of evidence on Climate Change and its economic costs. Its findings were clear: if we let Climate Change run unabated then it will bring far greater economic issues than if we don't act now.

Before we carry on, let's define one term, namely the GDP, or Gross Domestic Product. This is the monetary value of all the goods and services.

The Stern Review suggests that we could mitigate Climate Change by annually using just one percent of the world's annual GDP. Now, this is still a lot of money. In 2018 the world's GDP was eighty-six trillion dollars[42], so one percent of this would be eight point six trillion dollars. A lot of money! But given that, pre-covid, the world's GDP was increasing by an average of three-point-five percent each year[43], then maybe it's not that much. For some context, as mentioned in Chapter Three, the fossil fuel industry received way over half of this amount in subsidies alone!

The Stern Review also stated that the longer we leave it, the more expensive it will be. Ignoring Climate Change completely could, by as early as 2050, cause economic damage of up to twenty percent of the world GDP! Therefore, even if we don't really care about people, biodiversity, and a whole host of other issues, no, even if we only really care about money, then it still makes absolute sense to do all we can to mitigate Climate Change now. Only those who care more about themselves right now wouldn't want to make that change occur.

Note that a GDP growth rate of three-point-five percent each year leads to the GDP doubling in just twenty years! A growth rate of two percent would lead to doubling within thirty-five years. Whereas a growth rate of one percent, well below where modern societies strive for, would still see the GDP doubling within seventy years. The Earth can't even sustain what we are producing now, let alone more!

We will leave this section with several thoughts;

When making decisions and carrying out actions, how much responsibility do we have to others around the world and to future generations? Just imagine if you'd been born into a world already at two, three, four, or even more degrees Celsius of warming above pre-industrial levels. How would that make you feel about past generations, who knew the solutions but chose to do nothing?

Now consider how the wealth of the planet is shared out. The wealthiest twenty-six people on Earth own as much as the poorest fifty percent[44]! In what way is this fair or acceptable? It has been said that the richest deserve their money because they work hard for it. Tell that to those who work several different jobs just to keep a roof over their head. The current system is effectively telling us that some people's time is worth more than others, and by extension, those people are more important as human beings than others. Is this really ok?

Imagine a world in which those organisations, such as corporations and governments, which are doing the harm, had a legal requirement to deal with and fix it. Imagine that this came with, not just a fine for not complying, but with CEOs (Chief Executive Officers) and the like being under threat of prosecution, including jail time, for not complying. Imagine that polluting corporations had to spend their money cleaning it all up from our environment. Imagine that those who produce all the plastic were responsible for its clean up. I believe that the world would be a very different place indeed.

Since huge subsidies, paid for with our taxes, are being given to the likes of the fossil fuel and animal agriculture industries, this essentially means that we are literally paying to create harm in our world! How can this system make sense? Also, many financial institutions, such as banks, and pension schemes, put your hard-earned money into the likes of the fossil fuel industry to make themselves profits. You couldn't make it up!

But there is still hope. There have been times when governments have come together for good. National and even international legislation has been created and implemented to bring about some positive change. Let's have a look at this now.

Legislation

Legislating is the making and implementing of laws. Not all this section is actually legislation, however. Sometimes worldwide agreements are made but are not legally binding. UN resolutions, for example, are not generally legally binding, but they can be (note 'can be', not 'are') used to exert pressure on governments who aren't acting within them.

One prime example of successful global collaboration is the Montreal Protocol. In the 1970s, it was recognised that CFCs (chlorofluorocarbons), which were used extensively within refrigerators, air conditioning and spray deodorant, were damaging the ozone layer[45]. The ozone layer is a region high up in our atmosphere that contains relatively high amount of ozone, which is three oxygen atoms joined together. This ozone layer is very effective at absorbing the damaging, high energy, ultraviolet rays from the Sun. Without this layer of ozone, the ultraviolet would cause so much damage that, without it, many plants and animals could not exist on land.

It was discovered that the CFCs could end up in the upper atmosphere, break down, and release chlorine. This would then react with the ozone, setting off a chain reaction, causing each molecule of CFC to cause the loss of thousands of ozone molecules. This was making the ozone layer thinner, most drastically over Antarctica, leading to the so-called hole in the ozone layer.

Many who did not want to act upon this evidence argued against the science, trying to exploit the inherent uncertainty, which is an integral part of all science, along with the complexities within our atmosphere. Then in 1984, studies showed overwhelming evidence that the ozone layer had decreased significantly since the 1960s. Policy makers could no longer go against the science.

In 1987 the Montreal Protocol on Substances that Deplete the Ozone Layer was signed, and came into force from 1989. Since then, there have been six more amendments, with the latest being the Kigali Agreement in 2016[46]. The outcome of this is that the ozone layer is recovering by between one and three percent each year[47] and is expected to have made a full recovery by the middle of this century.

The Montreal Protocol is to be held up as a shining example of what we can do when we have strong scientific evidence on what is happening, why it is happening, and what we can do about it. We are in a similar position on Climate Change. The difference is the scale, and the political will. The changes required are far greater than those regarding the ozone layer, but if we come together, as was done before, then we can succeed.

As seen earlier in this chapter, scientists have understood the effects of burning fossil fuels, and of increasing concentrations of greenhouse gases, like carbon dioxide, for a long time. We have very clear, well understood scientific evidence on what is happening regarding Climate Change, why it is happening, and what we can do about it. Let's then look at what has happened on the global stage.

The first major global body to work on this was the IPCC (Intergovernmental Panel on Climate Change) which was established by the United Nations Environment Programme (UNEP) and the World Meteorological Organisation (WMO) in 1988. The aim of the IPCC is to 'provide policymakers with regular scientific assessments on Climate Change, its implications and potential future risks, as well as to put forward adaptation and mitigation options'[48].

The IPCC produces 'comprehensive Assessment Reports about knowledge on Climate Change, its causes, potential impacts and response options. The IPCC also produces Special Reports, which are an assessment on a specific issue and Methodology Reports, which provide practical guidelines for the preparation of greenhouse gas inventories[49].' To date there have been six Assessment Reports, with the latest having been published in August 2021. There have also been many Special and Methodology Reports.

For each of the reports, a comprehensive group of scientists from across the globe are brought together whom 'reflect a range of scientific, technical and socio-economic views and backgrounds'[50]. The Sixth Assessment Report, for example, had seven hundred and forty-three experts. The IPCC does not conduct its own research, but instead uses many thousands of peer-reviewed scientific, technical, and socio-economic papers. A first and then second draft is created, and then reviewed, before the final report is published. This is an extremely rigorous process and based solely upon the known research at the time. It is worthy of regard and to be respected greatly regarding its findings and use.

Some of the key findings from the Sixth Assessment Report are as follows[51];

- the Earth's climate is changing, and the role of human influence on the climate system is undisputed
- carbon dioxide (CO_2) is the main driver of Climate Change
- many of the changes observed in the climate are unprecedented
- some of the changes are irreversible over hundreds to thousands of years
- unless there are immediate, rapid, and large-scale reductions in greenhouse gas emissions, limiting warming to close to 1.5°C or even 2°C will be beyond reach
- global temperature is expected to reach or exceed 1.5°C of warming over the next twenty years
- warming over land is larger than the global average, and it is more than twice as high in the Arctic
- Climate Change is already affecting every region on Earth, in multiple ways

- with 1.5°C of global warming, there will be increasing heat waves, longer warm seasons, and shorter cold seasons
- with 2°C of global warming, heat extremes would more often reach critical tolerance thresholds for agriculture and health
- more intense rainfall and associated flooding, as well as more intense drought in many regions
- coastal areas will see continued sea level rise throughout the 21st century, contributing to more frequent and severe coastal flooding
- further warming will amplify permafrost thawing, and the loss of seasonal snow cover, melting of glaciers and ice sheets, and loss of summer Arctic sea-ice

Clearly, this is not pleasant reading and further amplifies the importance of systematic global action now.

There have been many other reports, conventions and protocols over the years, and there is no need to mention all of them here, but let's look at a few of them in detail. Those mentioned below are all part of COP, i.e. the Conference of Parties. COP is an annual event, bringing countries together from around the world, with the first one taking place in Berlin, Germany in 1995. In many ways these are amazing in that most countries across the world get together and discuss necessary action on Climate Change. The very fact that it recognised to be of enough importance that almost everywhere is represented and committing to action should be celebrated. Let's see some of their actions so far.

In 1997 at COP3 in Japan, the Kyoto Protocol[52] was adopted. This commits countries to adopt their own policies to mitigate Climate Change, as well as reporting periodically on their actions and findings. Recognising that the 'high-income' countries are largely responsible for the current high levels of carbon dioxide in the atmosphere, a greater stress is, correctly, put on their actions. Initially these countries committed to reduce their carbon emissions by five percent from 1990 levels. A second commitment was then to reduce emissions by at least eighteen percent by 2020.

Whilst the aim was to genuinely reduce their emissions, there was also included a mechanism known as offsetting, in which countries could 'buy' reductions in emissions from other countries. This meant that countries could continue to emit at high levels, but then pay better for the reduction in emissions of others. They would still be deemed to have met their targets!

In 2015 at COP21, the legally binding treaty Paris Agreement[53] was adopted with a goal to limit Global Warming to preferably to no more than one-point-five degrees Celsius, and certainly less than two degrees Celsius. Within this, countries had five years to set, and share, their own Nationally Determined Contributions

(NDCs). Just to clarify that, each country determines which actions it takes to cut their carbon emissions.

An important part of this, the Green Climate Fund[54] was agreed, in which wealthier countries, who have caused the greatest emissions, would provide financial support to those less wealthy, to support their climate action. The full one hundred billion dollars agreed by 2020, was not, however, provided. And it is not just in payments in which not enough is happening. To date, as seen in the Sixth Assessment Report, not enough is yet being done to curb significant warming, and far great commitment and action is required.

This was clarified in the 2018 IPCC Special Report on Global Warming of 1.5 °C[55]. It was in this report that the 'Twelve years left' came from. It stated that, by 2030, carbon emissions would need to reduce by about forty-five percent from 2010 levels to have any real chance of reducing warming to one-point-five degrees Celsius. In other words, a very significant change in our emissions would need to take place. It never meant that we didn't need to act for twelve years, nor did it mean that we should give up if we don't meet this target in that time. It does, however, emphasise the need for concerted climate action right now.

A need for net zero by 2050 was also stated. That is, by 2050, there would be no more greenhouse gases entering the atmosphere than are being removed from it. Three things on this. Firstly, there are many different pathways to net zero. We could reduce emissions very slowly to begin with and then make lots of changes just before 2050 or make significant changes now. Whilst both methods, and any in between, would hit net zero by 2050, there are huge differences in how much carbon is emitted in this time. The science on this is so clear; the greater the emissions, the worse it is going to be, so surely, we have a responsibility to carry out serious action right now!

Secondly, whilst the concept of net zero is important, it should be noted that there are some who are using it to continue to pollute as usual, or even increase their emissions. They 'justify' this by stating that they'll be carrying out actions to compensate in other ways[56]. This often relies on the technologies, that as we saw in Chapter Six, are nowhere near as capable or reliable as stated. What we need is to substantially reduce emissions, starting now.

Thirdly, whilst net zero gets so much attention, this doesn't account for the significant amount of greenhouse gases already in the atmosphere. Even going to net zero would be nowhere near enough. With the amounts in our atmosphere, even now, the Earth would continue to increase in temperature for many years. What is needed is 'net negative'. In other words, we need to be remove significantly more greenhouse gases than are being emitted in order to control warming.

Then there was COP26[57]. This took place in Glasgow in November 2021 and was hoped to be a great turning point in our global commitment on Climate Change. So, what happened?

Well, many world leaders, scientists, policy makers,

climate activists, indigenous groups, and representatives of the likes of the fossil fuel industry (!) descended upon Glasgow to try and make a difference. If you want to listen to just one speech, then I highly recommend that of Kenyan activist Elizabeth Wathuti[58].

There were a number of deals to make genuine improvements. These included reducing methane emissions by thirty percent by 2030 and also a commitment from over one hundred countries to end deforestation. Whilst these could be so much better, they are an improvement on what we had previously.

Now, we are unlikely to ever get to a situation in which those countries who have caused the most emissions will take full responsibility for their actions, but at least an increased financial package for poorer countries was agreed, providing one-hundred billion dollars per year to support climate action by 2025. When some countries have had so much financial gain by using huge amount of fossil fuels, how could it ever have been fair to stop others from doing so without providing them with suitable financial support so that they can develop without the need for fossil fuels?

There were also changes to the offsetting rules. Essentially, some of the loopholes in the previous agreement were closed so they could no longer be exploited. The problem was that these 'offsetting' actions were ones which needed to happen anyway, and so shouldn't have been able to be used to 'cancel out' emissions from another country.

There were calls to stop subsidies to the fossil fuel industry, and even switch some of this funding to the promotion of renewables. Unfortunately, not only are many in governments still lobbied heavily by the fossil fuel industry, but there were even a very large number of representatives from that sector present. In the end, the wording became 'phase out inefficient fossil fuel subsidies', whatever that means! At least there's a recognition that it needs to change, but clearly not enough has been done with this.

There were also calls from countries to reduce the use of fossil fuels. This should be a given. Certain countries, however, such as the UK, USA, and many in the EU, who mostly use gas and oil, wanted this to be just for coal, and not the other fossil fuels. Other countries, such as India and China, who rely more on coal than gas and oil, understandably felt that this was unfair and that it should be for all fossil fuels to keep a level playing field. In the end, it was agreed to 'phase down' unabated coal (note the word 'unabated'). As part of this, COP26 President Alok Sharma[59], along with a good deal of the media, put all emphasis on coal, and the blame on China and India, rather than why it was not agreed for all fossil fuels to be reduced. Still, unbelievably, it the first time any fossil fuel has been mentioned in any of these agreements, which says enough in itself, and it is great that coal use will reduce. It's just that this was an opportunity to do so much more.

One more thing to mention is the NDCs, which countries declared before COP26. Collectively, NDCs had still led to a projected warming of two-point-nine degrees Celsius. This latest round of NDCs have reduced this to two-point-four degrees Celsius. This was discussed and recognised to be still too high, so an agreement was made to further improve NDCs by COP27 in Sharm El Sheikh in Egypt. This is significant since previously, countries would have made these changes every five years, rather than one.

Finally, it should be noted that there were no agreements made on animal agriculture[60]. How can this be? In Chapter Four we clearly saw the importance of addressing this issue, so why was nothing done? Clearly fossil fuels cause by far the highest emissions, so had to be the main focus. But it has taken twenty-six COPs before any fossil fuel was even mentioned in an agreement. How long before animal agriculture is included? Thankfully, the UN has called for food to be central to discussions at COP27, but it is disappointing that they haven't already been included.

So, in summary, was COP26 a success? Well, certain countries seemed happy with it, alongside the fossil fuel industry, but many so-called developing countries felt let down, as did climate scientists. The truth is that it was an improvement on what we had previously but didn't go anywhere near far enough. We still have the situation in which certain 'global leaders' are not taking it seriously enough, and not taking responsibility for their actions. The concentration of greenhouse gases in the atmosphere is still increasing and, even with all current targets met, will do so long into the future.

But China, though

There are many who say that there is no point in carrying out our own actions, whilst the likes of China emit far more. This is the 'but China, though' argument, and has been used time and time again divert blame and not take responsibility for our own actions. So, let's have a look at this in more detail.

Well, when we look at the data[61], we see that China does indeed cause greater current greenhouse emissions than anyone else, so maybe they have a point. But China also has a far greater population so maybe a fairer way to look at it would be emissions per person. Unless we believe that people in some countries are less valuable than others, then surely this is fair.

For anyone else still in doubt whether 'per person' is a fairer way to look at it or not, then just imagine if China was split up into a number of different countries. Where would your argument be then, even though the total emissions would be the same?

Ok, so looking at emissions per person, we find that China is forty-sixth on the list[61]. How can we then go on blaming China? I live in the UK, and continue to hear the 'but, China though' argument a lot. So, let's compare the UK and China. According to this data, China, whilst forty-sixth on the list, does emit more greenhouse gases per person than in the UK. But is this the whole story?

Further inquiry leads us to the fact that not all emissions are included. Indeed, air travel, shipping, biomass, military, and even the production of goods and services that a country imports from another, are not included[62]! This means that a country can rely upon others to do their production of goods, otherwise act as they always have, and claim lower emissions.

When we look at the UK and China, we find that the UK is a significant net importer of such goods, whereas China is a net exporter[63]. Adjusting for imports and exports then, we see the UK move from five-point-five tonnes per person, to seven-point-seven-five, whereas China moves from six-point-seven-two to five-point-nine-nine. i.e., when we look at the full picture, the emissions per person are lower in China than in the UK.

Clearly then, those in the UK have little argument when it comes to blaming China for the world's emissions. Only those who believe that only the total should be looked at, essentially stating that each person should be allowed fewer emissions per person in China, has any argument at all. And there's a word for people who believe that people from some countries are more valuable than those from others.

It also seems fair, and reasonable, to look at historical emissions, rather than just those occurring now. After all, emissions from the past have very much contributed to our current situation. Here we then see that the USA has contributed almost twice as much carbon dioxide into the atmosphere as China, with about a quarter of all emissions[64].

Now, no one is saying that China's carbon emissions don't need to reduce. They do. But they also do in most places around the world. To apportion the blame just on China, is neither scientifically accurate, nor is it helpful. Indeed, pitting one country against another will never work, there is too much evidence of that in the past. We need to recognise, however, that some of us cause far greater emissions than others. So, is there a group of people who are doing the most harm?

The world's richest one percent, according to an Oxfam report, cause double the greenhouse gas emissions of the poorest fifty percent[65]. How is that for inequality! We also know that it is the poorer, more vulnerable that are affected most by Climate Change. If you aren't sure where you fit into the world in terms of wealth by the way, try a world wealth calculator, you might be surprised[66]. It gets even worse when we look at the super-rich. With a combination of private jets, journeys to space, huge houses and estates, yachts, and other factors which go with such a decadent lifestyle, a group of twenty billionaires were found to have average carbon emissions of around eight thousand tonnes per year[67]. Compare this to the global average of less than five. One billionaire, who Chelsea football fans will know well, even reaches over thirty thousand tonnes per year! That's over six thousand times more than the average, and about a million times more than the average person from the Democratic Republic of Congo. So, if there is anyone you really need to blame, then don't make it about this or that country, but look at the individuals, industries, and systems that support them, which are really doing the harm.

Since we are currently looking at emissions, let's also look at how much of all emissions have occurred within your lifetime[68]. It is found that if you are fifteen years old, around thirty percent of all greenhouse gas emissions have occurred since the day you were born! For a twenty-five-year-old it's more like forty-five percent, and for a fifty-year-old, it's about seventy-five percent. These numbers give a clear indication of what has been happening in the recent past, even though we have known so much about the damage it is causing and will continue to cause.

Within this, we do need to recognise that there are countries, for example many of those in Africa, who have contributed, and continue to contribute, very little in terms of emissions. Not only that, but Africa, despite having caused such a tiny percentage of all emissions, is feeling devastating effects from Climate Change[69]. So instead, rather than more 'them-and-us' and detaching ourselves from our own responsibility, let us fight to change a system that has climate destruction built into it, take responsibility for our own actions, and provide the support for others who need it most.

We can't even just put all blame on the super-rich. By attacking others, getting angry with them, and just telling them that they are wrong, their family and friends are wrong, their understanding of life is wrong, they will trench in and fight back, increasing their convictions and their standpoint. No, it is through working together and helping each other that we can achieve real change. Most people don't want to be seen as a pariah, and with awareness and understanding, positive change has a greater chance of occurring. In many ways, they are just parts of the system that is in such dire need of a change.

We have seen that there are significant variations in emissions across countries, industries, and individuals, but who is already being affected most? As usual, it is the poorest, the most vulnerable who are more at risk. Whilst they have caused the least harm, they just don't have the resources to adapt to the change that is already happening. On top of that, many such people live in the regions of the world which are already feeling the effects of Climate Change the greatest.

Africa has been recognised as the continent most vulnerable to the impacts of Climate Change[70], despite accounting for less than three percent of all carbon emissions. The continent has already seen increased drought and floods, affecting food production, water availability, disease, biodiversity and even climate refugees. Indeed, it has been estimated that by 2050, one hundred-and-five million people will be forced to leave their homes because of Climate Change[71]! For context, that's over one and a half times the number of people in the UK, or an amount which if all from one country, would be the fourteenth most populous in the world. In 2012 alone, over six million people in Nigeria were forced to leave their homes due to floods, with a further half a million in Chad[72]. How is this ok? This is real and happening to people right now.

This has been documented within the 2014 IPCC Synthesis Report[73] which states the following effects of Climate Change around the world; Africa is most at risk from reduced water availability, crop failure, reduced food security, and an increase in disease. Asia is most at risk from both drought and flooding, reduced water availability and food security, heat-related deaths, and damage to infrastructure. South and Central America are most at risk from flooding and reduced water availability, reduced food production, and increased spread of disease. Oceania is most at risk from damage to coral reefs, increased flooding and damage to coastal regions. North America is most at risk from increased wildfires, heat-related deaths, and flooding. Europe is most at risk from flooding and an increase in extreme heat events and wildfires. Last, but certainly not least, small islands are identified to be at most risk from the effects of increasing sea levels, loss of land, infrastructure, livelihoods, and ecosystems.

Clearly, everywhere across the world will see, and in some cases are already seeing, the devastating effects of Climate Change. None of us are invulnerable to this, though the effect on some, especially those less well off financially, will be far greater than others[74].

We must also not overlook the role of indigenous groups across the world. So often they are on the front line of the damage being done. Just look at the effects which deforestation[75], mining, pollution and other atrocities have had on indigenous populations around the world. Damage caused by a system hellbent on profits before all else. And what do we do about it? The media hardly report it, and even when we do learn about it, we generally either choose to do nothing or feel completely helpless to do anything. But these are people who live, not separate from, but as part of nature, with a great level of equity amongst the people.

We have so much to learn. No climate action can be truly global and representative without significant contributions from indigenous people.

Population

Some say though that there is no point doing anything since our world population is too high. Our global population has indeed increased greatly[76]. Back in 1804, there were one billion people on Earth. By 1974 it was four billion, and today we are at around seven-point-nine billion. By 2050 we are expected to be at nine-point-seven billion. That's a lot of people. The growth rate has halved in the last fifty years, but it is clearly still going up. So maybe population is the real problem. Let's have a look.

The damage we are doing to the Earth all relates to how much of the Earth's resources we are extracting, and the emissions we are causing. Surely, with a greater number of people, more resources are required, and more emissions will be released. Well yes, but it is not this simple. Factors such as which country you live in and how much money you have, make a huge difference. The average person in the USA, for example, uses over seven times the total amount of Earth's resources as the average person in India[77]. The country with the most resources used per person in the world in Qatar, whereas the least is in the Occupied Palestinian Territories[78].

When we look at the data, we see that people in many countries do not use more resources that the Earth can sustain. This means that we could all live within the

Earth's means. It also means that, since some of these countries, such as India, have a high population, it is not the number of people, as such, that is the problem. No, the problem is the aforementioned political and economic systems, and how we treat the world. If everyone used as many resources per person as in the USA, then we would need the equivalent of five Earths just to be sustainable. For the average person the UK, it would be two-point-nine Earths, and in China two-point-two. If we lived like the average person in India, however, then we only need zero-point-seven Earths, in other words our planet would be more than enough. What's more, there are fourteen other countries in which the average person uses even fewer resources than in India. Again, what is clear from this is that we certainly could sustain the current population, and noticeably more, if we didn't exploit the world the way we do in the financially wealthier countries.

Another thing to note when people blame the rising population for our Environmental Crisis, is that they always seem to be blaming poorer people in other parts of the world. Again, there is a name for people who only want to blame people from other countries, from other religions, from other 'so-called' races (Note: race is a social construct, not a biological one). It should make us wonder though, which people alive today would they choose to not exist? If people suggest that the Earth's population should be half of what it is now, then maybe ask then to choose which half of their friends and family they would be happy to not exist. Maybe ask them to roll some dice, with their name included too. It's as if there are those who think that some people deserve to be alive, but others do not.

It should now be clear that we cannot simply blame our environmental emergency on our population. Of course, if everything else was equal, then each increase in population would lead to an increased pressure on resources, but this is not the case. When many, particularly in the financially poorer countries, increase in population, then this increase has very little or even negligible impact on Environmental Crisis. When the financially wealthier increase in population, however, this increase really does put extra pressure on an Earth already exploited beyond its limits.

Despite this, population, itself, continues to be blamed. What happens here is that it distracts from the key issue, which is the political and economic systems which have the plundering of the Earth as a key ingredient. As mentioned previously, those in positions of power would much rather us blame each other, and cause separation, rather than have us see the current systems for what they truly are. In other words, increasing populations can, and do, have an effect, but the extent of this depends greatly upon where they live and how much money they have.

Note that this is in no way suggesting that poorer people don't deserve all the genuine benefits which come with an increase in wealth, they absolutely do. It should be a matter of shame that there are such extremes in wealth, not just in different parts of the world but also within each region. No, the changes that need to occur must include a levelling of resources, such that the current outrageous disparity no longer exists. The super-rich and the desperately poor must be no more.

As part of this, we should look at why the world's population is increasing. Globally, birth rates are decreasing. The main reason then for the increase in population is because there has been an increase in life expectancy. In other words, people are living longer. This is due to amazing advances in the likes of food security and medicine. Improvements in water systems which both bring clean water, and remove waste, have also played a significant role. All these achievements are very much to be applauded. Let's then look at if there is anything than can be done within this regarding the population.

We cannot ignore the fact that population increases are greatest in the poorer parts of the world, particularly in those where girls have limited education. Women across the world want to be educated, and to choose when, and if, they become pregnant. UNICEF have determined that educating girls has a huge impact, not only upon their own lives, but also those of their communities[79]. When girls receive a secondary education, they are less likely to marry young, and more likely to live healthier lives. They earn more money, strengthen their community, and reduce inequality. It not only improves their own lives, but also the increases opportunity for boys too. The longer girls stay in education, the fewer, yet healthier children they have, along with an increased choice over when they have them.

Sadly, this is not so easy to achieve in many parts of the world, and around one hundred and twenty-nine million girls do not attend school. This is for a variety of reasons, including poverty, child marriage and gender violence, as well as boys being prioritised over girls. All this means that, if you care about the lives of others around the world, or just want to see a reduction in population growth, then supporting girls' education is probably the best way to do it. All areas also need free, local family planning services as well as free contraception, which is in keeping with cultural practices and beliefs.

In any section on population, we need to also look at the movement of people around the world. Some areas of the world are already becoming too difficult to live in. Heat waves, floods, drought, cyclones, and these combining to reduce food and water security, are having a devastating impact on communities across the world. Of course, this has happened in the past, but we have seen that this is increasing, and will further increase in the future.

What would you do if faced with such circumstances, when the region in which your family have lived in for generations can no longer support human life? Desperate to support your family, you would move to search for a better life, or indeed any life at all, and rightly so. Whilst often called economic migrants, they have been forced away from their land by the effects of Climate Change. Going back would put their lives in danger. So, whilst there is no official definition of a climate refugee, and Climate Change is still not listed in the Refugee Convention[80], and a good deal of such displacement is within the same country, they should be given all the rights applicable to those who have been forced to move due to persecution.

In some parts of the world there is a lot of resentment towards those who move into an area. In many cases these same people are those who either deny Climate Change[81], or at least play it down and use arguments such as the 'but, China though' one seen earlier. So, the very people who don't want more people coming into 'their country', are also often the ones who don't want to do anything to reduce people becoming refugees in the first place. This Refugee Convention mentioned above also makes it clear that refugees do not have to seek refuge in the first country they come to. This is a myth, and rightly so.

There were over twenty-six million refugees in the world by the end of 2020[82]. Climate Change may not yet cause war directly, but it certainly makes conflict more likely, and makes adapting to that conflict far more difficult[83]. If ever in doubt whether you would happily accept refugees where you live, try to learn about the situation in which they come from and the causes of that. What would you have done if you were in their position?

War itself is also extremely damaging for the planet, beyond the obvious destruction of human life. Militaries own vast areas of land and use a huge number of resources. The military of the USA, by itself, contributes more greenhouse gases than most countries[84], whilst the annual emissions for the world's military add up to a well over three billion tonnes[85], which is more than all aviation, shipping, and waste put together[86].

War also damages ecosystems and creates pollution, especially during conflicts, where explosive weapons pollute air and soil, as well as destroy ecosystems and infrastructure[87]. It also diverts funds and resources away from where they could be otherwise used. There is little doubt that wars bring devastating effects to people and environments, even when supposedly done to protect others. Consider this; if the principal aim of the military is to protect the people, then doing all they can to reduce our Environmental Crisis would go a long way towards it.

The whole idea of wars and why they occur could be, and has been, the subject of entire books, but be under no illusion; there are those who make huge profits from war. And these very same corporations which make such huge profits from selling arms all over the world[88], are also significant lobbyers of government[89]. This is yet another example within our political systems that allow hugely powerful companies to influence policy, to increase their own profits, whilst causing great harm for others.

The worst thing about this though, is that whilst we fail to see each other as one, upon a planet on which areas are ever more damaged by the effects of Climate Change, then war will become increasingly likely. As food and water security reduce, desperation will lead to an ever-increasing number of people moving from their homes to other places. Some will resent this movement, and the fight for resources will become greater than ever before. The need to understand that we are one, and that diversity within this only makes us stronger, will become more important than ever.

Our Health

There is little doubt that the Environmental Crisis in which we find ourselves is having a variety of significant negative impacts upon human health. Let us look at some of these in turn.

Some such impacts are more obvious, such as those from heat waves, flooding, drought, and cyclones. But these all bring a loss of food and water security, so vital to human life. Each of these are getting worse, and so doing all we can to mitigate Climate Change is then also a fight for human health.

Then there is disease. Not only does a lack of water and food security make people more susceptible to disease, but there are more direct impacts, such as the increase in waterborne diseases from flooding[90], and the increasing areas in which the likes of malaria[91], can spread into.

We also saw in Chapter Four that our current practices, including eating meat, trading wildlife and deforestation greatly increase the chances of zoonotic diseases[92]. This has been felt acutely in recent times with the global pandemic of SARS-CoV-2, or covid. There is no doubt that this has had a devastating impact in terms of human life, as well as other health issues, including mental health. It has also had a significant impact on global economies. So, which caused it?

Well, whilst is it very difficult to prove, a WHO report[93] shows that the most likely cause was the transmission of the virus, from either bats or pangolins, to farmed animals, where the virus further evolved before passing onto humans. Other possibilities, such as direct transmission from bats or pangolins to humans, and directly eating animal-based foods are considered possible but less likely, and an outbreak from a lab incident is considered extremely unlikely. In other words, it is extremely likely that human actions, like encroaching on wild animal land, and farming livestock, made this happen. Indeed, it was even predicted years ago[94], that with our current practices, such a pandemic would emerge. If we wish to reduce the chances of similar occurrences in the future, then our actions, and the way we treat our planet, need to change.

If there is one positive to come out of this, though far outweighed by the negatives, it's that it has shown us that we can change. We can all learn and adapt our behaviour to make things better for us all. We know that taking over wild land puts animals under stress, leading to higher viral loads, and so they are more likely to pass them on. We also know that farmed animals are responsible, whether as origins, or transmission hosts, of zoonotic diseases. So, what are you going to do? What positive changes will you make in the food you eat, the clothes you wear, and the products you buy? Do you want human actions to increase, or decrease, the chances of similar zoonotic diseases in the future?

Then there are the issues with plastics. We know that we eat, drink, and even breathe in microplastics every day. We saw in Chapter Five that the average person ingests a credit card worth of plastic every week. These plastics have been associated with inflammation, cancers, and infertility. They can also attract micro-organisms, such as bacteria, which can then enter our bodies on the plastic[95].

If all this wasn't reason enough, a study has shown that chemicals known as phthalates, which are used in manufacturing plastics, are causing a growing number of babies to be born with small penises[96]. There are also observed effects on male and female reproduction, as well as behaviours.

We must also not forget climate anxiety. This is when Climate Change, and its effects, cause fear, stress, and worry. Around sixty percent of young people are found to be very worried or extremely worried about the effects of Climate Change[97]. So, what do we do about it? Well, we could either try to shelter people from the truth, or we could empower them so that they can both make a positive difference and have opportunities to express their feelings, ideas, and actions. I know which one I would choose.

Education

As Nelson Mandela once said, 'Education is the most powerful weapon which you can use to change the world'[98]. Since change is clearly what is needed, then education has a key role to play. If people don't understand the situation and what they can do about it, then how can we ever expect positive change to occur? And this applies to all levels of education. There are many highly educated people who know very little about Climate Change and the rest of our Environmental Crisis.

One of the problems is that, whilst most people are aware that Climate Change and other environmental issues exist, not everyone has a real understanding on what the issues, causes, consequences, and practices to reduce the harm, really are. And why would they? People have busy lives, with their own worries. Students in school are told just to focus on their grades. Many others don't want to come home from a hard days' work and learn about the Environmental Crisis we are facing. Those looking after families have other priorities. And many others spend their time trying to get work and find ways to cope with their lives. Then there all the distractions, such as TV. It is often much nicer to do other activities rather than concern ourselves with the bigger world issues.

We can genuinely blame no individual for not knowing more, unless perhaps it is part of their job to do so. Even for those who want to learn more, there is so much information, as well as misinformation, whether deliberate or otherwise, that it can be a minefield for someone trying to pick their way through it. That is one of the reasons for writing this book, in hope that it may help some people at least.

The problem is though, we are facing a Climate Emergency. We have seen that not enough is being done. Pressure and change must come from the people. But, until enough of us really understand the issues then how can we be expected to push for real change? We may know enough to carry out small, individual actions but the magnitude of change required will only come from changes in the system. How can we make this happen if we are kept in the dark, whether directly or indirectly?

Many people also have such faith in our political system that they believe that if the issues were really as bad as climate activists state, then the politicians would do something about it. Sadly, this is clearly not the case. Indeed, in many cases, the policy makers do not really understand the issues themselves. This includes even those politicians in charge of environmental policy having very little understanding of the real issues[99].

What we need instead, is an increase in awareness of the truth regarding our Environmental Crisis, and what we can do about it. The actual truth. This can come in many forms, and the aforementioned media and politicians have a real responsibility to deliver. Something that, quite frankly, they are failing at spectacularly. And we have seen why. So, what are we to do? Well, there are many ways in which we can both learn more ourselves and support the learning of others.

Let's start by looking at schools. Whilst change needs to happen now, before many young people are able to directly influence policy, supporting the learning of our young people is a key factor moving forward. Students in all schools, in all nations, would benefit from a strongly embedded environmental curriculum. This doesn't mean separate lessons on Climate Change and the rest; these issues are not separate from our actions in life, so neither should be our learning about them. There isn't a single subject of study, or specific job, which doesn't have a connection to the Environmental Crisis. Therefore, every subject needs to play its part. And what's more, students want to learn about Climate Change and the other issues[100]. Surely then, as a society, we have a moral duty to support this.

Through different subjects, students should be able to learn about, investigate, study, be creative with, explore and express their feelings on, the various aspects of our Environmental Crisis. We know that just learning the facts isn't enough. If this only occurs in Science and Geography, where will students share their emotions, their frustrations, their thoughts? Where will they get to debate, and question, and test their own understanding of these issues? Where will they get to express their feelings creatively? Students can then bring in ideas from one subject to another, further recognising that it is, as we all are, part of a whole. It is only when there is a well-developed curriculum, in which students have the opportunity to push their own learning, that true understanding will really come. And let's be clear, this isn't about telling the students what to think. It's allowing students to see the evidence, look at the research, understand how it has been developed and by whom, and then challenge their own thinking, as well as that of others, through argument, discussion, debate, and creativity. It's allowing the students to develop their own, evidence-based thinking, alongside opportunities to really express themselves.

What we know is that being told the facts alone is not enough for many people. It's unpleasant, uncomfortable information, and whilst we may hear it, we do not always really take it in. To do so can induce stress and anxiety, which can lead to despair and cause people to lose hope. But the alternative, in not allowing them to develop the skills, tools and understanding to both adapt to, and mitigate against the Environmental Crisis, is far, far worse. That is why it is so important that as part of this environmental curriculum, students have safe places in which they truly explore their feelings, be creative, and share. It is vital that each student knows that they are not alone in this, and that, together, we really can bring about positive change.

We also mentioned previously that education can bring so much to a community, especially in educating girls in poorer parts of the world. A bona fide climate curriculum, adapted to suit different student needs, has the chance to genuinely reduce inequality. The truth is that education should be a great leveller; it should provide equal opportunities to all. We know, however, that it does not. Neither will it under the current political and economic systems. Disparities in levels of education perpetuate the injustices within society. If we all get to really explore the truth, however, then maybe we have a chance.

The current education system, sadly, does not provide students with the aforementioned opportunities. Whilst countries vary in just how much is built into the national curricula, all too often there is very little. When present, it is often just within Science and Geography, which, as we have seen, is nowhere near enough. There are those, such as 'Teach The Future'[101], who are pushing for the required change, but presently too little is implemented in schools. There are also many organisations which provide, often free, resources on teaching these issues, but until it is fully implemented into our curricula, it will often be little more than tokenistic.

Even where it is taught, the science is sometimes inaccurate. We saw in Chapter Three that it is not an over-simplistic view that biomass is carbon neutral, but that it is quite wrong. Just recently, whilst marking a 2020 GCSE Physics paper, I noticed that the mark scheme did not allow students to state that biomass releases carbon dioxide, since it is (supposedly) carbon neutral. A cynic may conclude that this is all part of propaganda to convince students of something that is not correct. We could also, just as easily, interpret this as being a different issue, in which the examiners themselves haven't understood the intricacies of using biomass as a fuel, and made a genuine mistake. Either way, it emphasises a point, namely that it is not only the students who need to learn, but the exam boards and the teachers themselves. How can we support students with issues that we know so little about ourselves?

There is also a need to take care with the narrative regarding environmental action in schools. I have seen, first-hand, that when first introduced the students, who so often feel a need to have a level of rebellion in their lives, are extremely enthusiastic and so many get involved. When such action becomes the norm within a school though, it may be perceived that it is no longer the uprising that they first felt it to be. If students want rebellion, then who should their focus really be upon; those teachers who care for them and are trying to support them, or those governments and corporations who cause them harm?

One more aspect that needs to be considered within education is how it currently, deliberately, pits students against each other. Reinforcing time and time again that they have to win against each other to succeed. That by working hard we get to beat more and more people. It's an inbuilt form of elitism, which makes students feel superior, or inferior, to their fellow students. It creates a 'them and us'. The actual reasons why some students achieve higher grades than others are many, and intricate, with natural ability only playing a supporting role. Our Environmental Crisis is too big for all of this. Within celebrating diversity, we need to recognise the importance of working together, with a common aim of making the world a better place for all.

But it can't just be through schools that we learn. As mentioned previously, the media have a huge responsibility to share the truth. Even where this happens, it can be difficult to decipher from all the misinformation, deliberate or otherwise, that exists. For those who feel able to, looking at the actual research papers and IPCC reports in a good place to find out the latest research. It would also be beneficial to join an environmental group, whether online, or in person, where you can discuss such issues, and really delve into how you feel about it and what you want done about it.

There are many other ways too, of sharing awareness. Anyone who has ever doubted the power of sport to bring about positive change, need only look at how Nelson Mandela used the influence of rugby in South Africa to help bring the country together[102]. And so many people love their sport, their teams, and the individuals who participate. They are role models, whether they like it or not, and their actions have a direct impact upon how others view the world. When football teams take short domestic flights to a match, despite the environmental impact, it sends a message to their fans[103]. Likewise, when teams, and individuals, refuse to carry out actions which are detrimental to the environment, and spread awareness of such issues, then their positive influence will be palpable[104].

Amongst the many other ways in which awareness is being raised about our Environmental Crisis, perhaps the most controversial is protests, or marches. Let us now then look at why people protest and the effect it can have.

Power of Protest

The rise of groups such as Extinction Rebellion, and individuals such as Greta Thunberg, has, no matter how you feel about them, done a huge amount to raise awareness on environmental issues. Their actions are a beacon of hope for some and focus of frustrations for others. Putting these aside for now, let's look at what history tells us about the influence of protests, and what they can achieve.

What we can see is that fairer taxes, workplace safety,

the abolition of slavery, an increase in civil and indeed human rights, women's suffrage, the fall of the Berlin wall, countries gaining independence, LGBTQ+ rights, the downfall of monarchies, and the ending of wars, have all been significantly supported by people standing up for their rights, when others wouldn't, through the power of protest[105]. Time and time again, positive change has come about from large scale, collective action which says NO! to those in power. So many of the rights that we take for granted today were deemed as thoughts of extremists in the past. That change came about through a coming together of individuals, making it clear through their words and their actions, that 'Enough is Enough!'

So collective protest really does have the capacity for change. In times when those in power are refusing to act upon what needs to be done, then protest is one way in which the rest of us can make a difference. But collective protest brings so much more. It is easy to feel isolated, and unheard, in such difficult times, as if no one else understands. This can lead us to lose hope. That is why it is so important that people have opportunities to come together, discuss, share, and support. By spending time with others who also feel the same way, we know that we are not alone, we feel valued, and that maybe, just maybe, there is hope after all. In many ways collective action is a form of therapy for those who need it.

So, when those 'in charge' are not doing what is right, then we have, not only the right, but the responsibility to stand up for what we believe in. There are many ways in which this can be done, but whatever form they take, it is important that they raise awareness across society, put pressure on those in power, and strive to be inclusive and bring others along with them. This includes the notion that protest must be non-violent, in part at least because violence goes against everything that most protestors believe in. Any use of violence creates legitimate reasons to be against it and can cause the protestors to be seen as those with lower morals, removing support from those who would otherwise be on their side.

Now, let's be clear, you aren't going to please everyone. For example, the mass protests, such as those from Extinction Rebellion and School Climate Strikes, which have included road closures, do annoy some people. They have, however, played a significant role in raising awareness and applying pressure. An integral part of protest is to make things inconvenient for those governments and corporations who are simply not doing enough. Whilst ideally, protests would just cause such inconvenience for those organisations, often their areas of work are so well protected that they are more difficult to target. Also, the media may ignore such events, thus removing the vital aspect of raising awareness in the general population. Saying that, many protests, have indeed targeted government buildings, fossil fuel companies, financial institutions, media, and the like, to primarily impact those. It is also worth considering that those who complain most about disruptive protests often take little account of how 'disruptive' our Environmental Crisis is for so many and that it will just get worse if we allow it to.

Obviously, since these protests are done to protect

people and the rest of the world, ambulances should be able to get where they need to go as quickly as possible. I have heard people complain that ambulances have been delayed, and this is something no one wants, but far too often these are the same people whose policies in government have had a disastrous effect on medical care, leading to ambulances waiting, with gravely ill patients, for many hours outside hospitals[106]. It makes you wonder what their priorities really are.

Also, climate protests are anything but an attack on society. They may well apply pressure for governments and systems to change, but this is done to bring about positive change for society, not to cause it harm. The people I have spoken to and been with on climate protests genuinely care. They are willing to be involved such protests and give up time in which they could have been doing other things, because they recognise the importance of such events, striving to make the world a better place. And it genuinely makes a difference.

Indeed, the increase in mass climate protests has applied such pressure, that one common tactic used is to discredit those involved. Any action carried out by protestors in their everyday lives, which isn't the best possible one in terms of solving our Environmental Crisis, is pounced upon. These accusations of hypocrisy include the fact that many of us still use fossil fuels within our lives, including their use to arrive at wherever the demonstration may be taking place. As a counter to this, consider that, within the present system, the best use of fossil fuels may well be to campaign against their use. As if anyone could be perfect anyway, particularly under the current systems.

Putting such judgement on others can make them doubt themselves, and not feel worthy of attending such protests or joining environmental groups. That is part of the reason why others do so; it makes those who haven't yet joined the movement less likely to do so. It can also make those who judge feel better about their own inaction.

Therefore, it is important that we always strive to include, and to be supportive of others, no matter how far they are from getting involved. We are all part of the same world and will all contribute to solving the issues. This very much includes yourself. Don't let anyone put you off. We are all on this journey together. As we learn, and make changes within our own lives, we must be mindful of the fact that it is not the same for everyone. We all have different understandings and capabilities to adjust our lives.

Another way, in countries like the UK, to try to remove the legitimacy of the protestors, is by inferring that they are all white and middle class. Firstly, for many people, there is too much going on in their lives to be able to protest. In this way, it may well be those more privileged who both choose to, and are able to be, activists. Secondly, it fails to recognise the numerous demonstrations and other actions from people across the world. In many cases they have had activism thrust upon them by how they are being treated and oppressed. This includes indigenous groups, who have been at the forefront of such protest long before many people took notice elsewhere in the world. Such inference also acts as another attempt to separate people into 'them and us'. I would be wary of anyone who encourages such separation.

Another criticism that some protesting groups receive is that they encourage working with the police. This is part of the recognition that they are people too, and thus need to be part of the solution. Not giving up on the those who are sometimes used by those in power to oppress others, is a key ingredient to making the world a better place. This doesn't mean agreeing with them, and it is important that they are called out when they carry out actions which do harm. But it does mean thinking of all of us as an integral part of the solution.

When protest continues to be legitimate and continues to raise awareness against the actions of those in power, then another resort is to make such demonstrations illegal. In the UK, for example, new laws are being drafted to give the police the power to stop any protest which causes road closures, or deemed to be too noisy, at the police's discretion. This basically gives the police the power to stop any protest at any time. The proposed laws would also allow the police to stop and search any protestor, without suspicion. How could such measures ever be ok in a just society? They would also be able to ban individuals from any protests in the future and include long prison sentences.

This should all make us wonder about those in charge, when they are more willing to bring in such draconian measures, at the expense of our human rights, than do the right thing for our world. They do so because they know that protest works, that it is a threat to their way of life. If it didn't, then why would the likes of BP pay for climate groups to be spied upon[107]? So, go forth and protest, be that change!

Making it Fairer

There is a difference between casting blame and taking responsibility for your actions. As part of reducing divisions, those who has benefitted in other ways from the doing the most harm, have a duty to clean up the mess they have made. Just imagine a system in which those who have cause the most harm, namely the richer individuals, corporations, and countries, used their money, power and resources to support others, and bring about positive change. Indeed, how could it ever be expected that those from lower incomes, whilst being the most effected by Climate Change, are those who have to carry out the greatest actions to mitigate it? Indeed, there are many people around the world with such a low income that they are doing all they can feed themselves and their families. How can we possibly expect those who have done the least harm to be the ones to change the most?

Many of those who have done the most harm, however, have shown little willingness to act in such a conscientious manner. For them to do so, legislation needs to change. For this to change, we must continue to put significant pressure on policy makers. If the movement became large enough such that no politician would ever gain a majority without such policies, then enough of them would make that change. Or more likely, other people would be voted in, who would carry out such actions. Likewise, if enough people refused to buy items from corporations, then they would soon have to change their ways to become more sustainable, and even restorative in nature. We can all apply such pressure. It is through the many actions in Chapter Nine, and a whole host of others, that we must hold these groups to account.

This must include, as mentioned previously, CEOs and others in high positions, being under threat of prosecution, including jail time, for not complying. It cannot be right to do such extensive harm and suffer no direct consequence. And for the rest of us trying to live our daily lives, remember that whilst our individual actions do make a difference, the system is such that, often, it is too hard to carry out the more environmentally friendly option.

We also need to change how governments provide very significant financial aid to highly damaging corporations. Currently, as well as the fossil fuel industries, many large multinationals receive vast subsidies, including those involved in airlines, motor vehicles, military arms, metals, clothing, and electronics[108]. All so that they can make bigger profits, allowing them to sell products at cheaper prices, meaning that we consume ever more. In other words, governments are giving enormous amounts of money, from our taxes, so that very rich people become richer, all whilst leading to greater damage to the environment. It just can't be the right thing to do.

Just consider a situation in which subsidies were instead transferred to positive actions. Imagine that they were instead used to both mitigate for, and adapt against, Climate Change. Imagine that energy subsidies switched from fossil fuels to the likes of solar, wind, geothermal, and ground source heating. Imagine that it was used to insulate homes and create a cheap, regular, reliable public transport service. Imagine that the money was used to reduce poverty, rather than further enriching the wealthy. Imagine that it was used to clean up plastic pollution. Imagine that it was used to increase biodiversity. Imagine the largest multinationals had to pay their fair share of taxes. With the fossil fuel industry alone receiving over five trillion dollars per year, on top of everything else, we are talking immense amounts of money. We could do so much positive with this.

Another way to make things fairer is the proposed Climate Income, proposed by the Citizen's Climate Lobby (CCL)[109]. In the Climate Income, whenever fossil fuels are extracted or imported, the business which does so has to pay a fee. Along with cutting subsidies, this makes the burning of fossil fuels more expensive. Consumers, whether individuals or in industry, will then switch to other, cheaper alternatives, which also happen to be cleaner. In other words, people will turn away from fossil fuels, rendering them obsolete.

Within this, a rise in energy prices from fossil fuels will occur. This can be reduced using subsidies for the non-fossil fuel energy resources. But CCL have a plan. The money collected from the fees would be a dividend, and distributed equally to each adult (half for children) within the country. Those on lower incomes, who generally use far less energy, should gain more money from the fees, than the extra they may have to pay in their energy bills. Those on higher incomes, who generally use far more energy, may well end up having higher energy bills, since the money from the fees won't cover all which they use. This essentially means that those who cause the most harm will have higher bills than previously, effectively paying extra for the damage they are doing. Not only does this seem fairer, but it may well make, some of them at least, look at how to reduce their consumption.

Green Careers

With so many potential opportunities to create a better world, let us now look at how jobs will play their part. Whilst there are many so-called green careers, which have a specific role in supporting the environment, in truth, every job should be considered to have an important role in bringing about a better future for us all. There is an environmental element to play in every profession. If we can just consider both the current impact of our work on Environmental Crisis, and how it can be adjusted and adapted to support positive change, then there is indeed hope for us all.

Within this, occupations may raise awareness, carry out direct climate action, or support the actions of others. Whilst some professions may have a greater impact than others, everyone has an important role to play as part of the whole.

Consider each of the following, which are in no particular order, and recognise the importance that they can play in bringing about positive change; waste collector, recycler, agricultural worker, energy technician, engineer, scientist, journalist, teacher, politician, emergency services, artist, musician, actor, architect, gardener, finance worker, manufacturer, salesperson, researcher, information technology expert, forester, food supplier, manager, business person, lawyer, military, medical staff, writer, and any other you can think of.

In short, whilst some professions may, indeed, be considered 'green careers', it is best to consider every career to be, or at least have the potential to be, environmentally friendly. Together, we really can make a difference.

Positive Change & Global Action

If we define positive change as anything that makes our world a better place, then how can that not be something that we want? We have also seen that it is when we remove barriers and see, not only all humans, but all life, as part of the same whole, that not only will we strive for greater such change, but our actions will have extra significance. Let us then look at a number of points on this.

Firstly, whilst there are a significant amount of greenhouse gases currently in our atmosphere, so things will get worse, we do have the power to control how much worse. Our actions count, the actions of every single one of us. We know that individual changes don't make as large a difference as systemic change, but don't discount yourself within this. You matter. We all do. Through this we can realise that all is not lost. We have an opportunity here to make a significant difference.

Secondly, it is important to realise that people across the world are already doing so much. Many individuals, and organisations, have made incredible positive changes both within their own actions, and through supporting those of others. An increasing number of people genuinely care about what is happening on our planet, future generations, and ultimately ourselves. You are not alone in this. Seek out those others and collaborate together in order to affect the kind of change we want to see in the world.

Thirdly, not making positive change is not a neutral standpoint. Changing nothing means that Climate Change and the Environmental Crisis, will continue to get worse. Not making change leads to inevitable ruinous consequences, not just for future generations, but for people and other organisms across the world right now. Are you ok with this?

Fourthly, consider the suggestion that climate action needs sacrifice. Whilst it may seem that way, and indeed we do need to adjust and reduce our consumption, a lack of such action will lead to far greater sacrifices in the future, including loss of land, food, and water for many. Instead, we could look at the positive changes as exactly that; not just positive for the Earth, and for others, but also for our own lives. With so much emphasis in today's world on having more, does it really lead to a happier life? Research tells us that the opposite is true[110]. Whilst gaining material items may bring short-term happiness, it is seen that our wellbeing actually reduces as we become more materialistic. Also, becoming less materialistic leads to an increase in wellbeing.

Therefore, if we know nothing about the environmental issues and just focussed on the welfare of people, then we would still conclude that we should reduce our desire for material goods.

As our pursuit of ever more material items reduces, along with other positive climate actions, then we will instead have more affordable energy, cleaner air, a greater sense of community, increased equality, more jobs, higher biodiversity, more beautiful places in locality, greater food security and less extremes of weather. I wonder now, does it really seem like a sacrifice to turn from materialism to positive action for all?

Summary

When we consider our actions, we should always contemplate, what type of world do we want to leave future generations? How would we feel about our ancestors leaving such a world for us? What is clear is that we can no longer consider the Earth as something separate from us, here to be exploited. Instead, we must consider ourselves, and indeed all things, as all part of a beautiful, intricate world. By living as part of nature, rather than against it, benefit will come to all.

The evidence tells us that our current systems are unfit to solve the Environmental Crisis. Those who presently make our policies, and inform our thinking, are all too often those who benefit most from maintaining a status quo and will use all manner of tactics to maintain their positions of power.

The whole narrative that 'we' are destroying our world is not fair. Those with the greatest money and power are doing the most harm. To suggest that population is the issue, whilst some individuals have a larger negative impact than millions of others, tells us so much about what the problems really are in the world.

But all is not lost. Carry hope in your heart. For an awakening is occurring all over the world. Despite all the deliberate misinformation, people all over the world are seeing the evidence for themselves. The issues are becoming ever more real for an increasing number of people. As public awareness of the reality of the situation increases, a tipping point in society will be reached, at which the necessary change will occur. What we know though, is that the sooner that change occurs, the less harm will already have been done, the less loss in biodiversity will have occurred, the less carbon will already be in our atmosphere, and therefore, the less dramatic the appropriate changes need to be.

Remember, this is our climate, our future, and there is a better way.

CHAPTER 9

SO, WHAT CAN I DO?

We have seen over the previous eight chapters that there are many critical environmental issues occurring around the world that need, not just our attention, but our action. To secure a healthy planet for all species, including ourselves, both now and in the future, positive change must occur. One way to summarise our requirements is with the following;

Awareness; that each of us genuinely understands the issues at hand, their consequences, and how to make significant, positive change.

Action; that we carry out the required actions, including global, regional, local and individual. This includes challenging systems, as well as global, local and even individual change.

Hope; that we recognise that there is a better way and that all positive change will make a difference. Through supporting each other and carrying our collective action we can create an improved world for all of us.

For the rest of this chapter, we will look at a range of positive actions that can support this change. The actions listed below are far from exhaustive. Please also use other resources, including each other, to help implement other positive changes. Even add your own actions, and feel free to let me know them, so we can support each other further, for the benefit of us all.

Just before we read through these actions, please remember that there has been, and continues to be, a deliberate and orchestrated spreading of misinformation from governments and large corporations. This comes especially from those whose profits will be impacted most from such change. Part of this is in trying to place the guilt at our door, suggesting that our own individual actions are to blame for the world's ills. This is, of course, not true. Whilst such actions do play a part, it is through large scale action, such as enforced policy change, that will have the greatest impact. This means that our most important actions are those which lead to real change regarding governments and corporations.

Also please consider that whatever the systems we currently have and the life each of us has lived so far, we can all start right here, right now, to further a better life for all. It is not too late for you. Join with others to bring about that positive change that you crave. And don't feel that you have to know everything or live a perfect life in order to get involved. None of us are perfect. If we waited until we are than no change would happen at all.

Also consider the actions that humans have carried out throughout history to fight for a better life. In that circumstance, would you have stood up to oppressive dictators, for the right to vote, in defiance of enemy invaders or joined the struggle against apartheid? Well, this is our war, our fight, and the consequences of not acting are all too severe. Will you be able to look in the eyes of future generations, and tell them that you did all you could?

It is also worth pointing out here that, whilst there are those who spread misinformation, the majority of people are actually good[1]. Most people want to make a difference, but everyone's circumstance is different. It is far more suitable for some people to make certain changes, than others. For example, no one is asking the Bajau tribe[2] of southeast Asia to stop fishing, but this doesn't mean that we should continue to consume the fish which is captured from industrial fishing. Likewise, most people cannot afford solar panels on their roofs, but that doesn't mean that those who can afford them shouldn't do so. Also, most of us are not in high-level policy making positions, but that doesn't mean that those who are, shouldn't do the right thing.

The actions below are split up into a variety of sections, though there is a huge amount of overlap between them. Rather than repeat myself, it seems sensible to write each action just once, with the understanding that a great many of these actions could fit into various sections. Also, whilst some actions will have more impact than others, those listed are not in order. Remember, every positive action makes a difference.

Use this list of two-hundred-and-sixty-five suggestions as suits you best. It may be that you wish to tick off or highlight actions that you do. I hope you find it useful…

Applying Pressure

1. Attend and even organise climate protests, to raise awareness, be with others, and put pressure on governments and corporations
2. Work with others to bring about systems which support a healthy planet, and therefore a healthy people, at their heart
3. Where suitable, carry out strikes, to put pressure on governments and corporations
4. Research the voting record and actions of those seeking election and share your findings on this with others
5. Vote for those who genuinely act, not just say they act, for positive change, and put the environment at the core of their actions
6. Write to your representatives, such as MPs and councilors on local, national and global action
7. Lobby government and representatives on;
    a. removing subsidies for fossil fuel companies, whilst increasing subsidies for solar and wind

b. signing up to, and putting into law, the fossil fuel non-proliferation treaty
c. implementing a carbon tax on all emissions
d. providing a Climate Income, such as that proposed by the Citizen's Climate Lobby
e. having those corporations who cause the environmental damage take responsibility for reversing that harm
f. removing subsidies for animal agriculture, whilst increasing subsidies for organic crops and rewilding
g. stopping and reversing deforestation across the world
h. setting up new marine conservation zones
i. supporting a national rewilding program
j. implementing a nationwide environmental curriculum, including all subjects
k. supporting less wealthy countries in their transition to more positive actions
l. legislating to stop water companies pumping sewage into our rivers
m. increasing affordable, electric, public transport

n. setting up low carbon zones within city centres

o. increasing the infrastructure for electric vehicles

p. dealing with waste, including recycling, in our own country rather than exporting overseas

q. making idling illegal, especially at hospitals and schools

r. having standardised labels on food produce about origin, freight method, whether produced in or out of season, and for how long the land has been used for agriculture

s. have stores such as supermarkets reduce food waste and packaging, as well as having doors on their fridges

8. Research the environmental and ethical records of different corporations and share this information with others

9. Write to those corporations on their actions and policies, and boycott those doing the most harm

10. Find out which pensions, banks and other similar institutions are financing the fossil fuel industry and similar, and share this with others

11. Switch pensions, bank accounts, and other financial dealing to those with strong ethical and environmental records

12. Write to the financial service companies causing environmental damage to express your concerns
13. Speak to sporting teams, bands, artists, etc, whom you support, about your concerns and how they can improve environmentally
14. Support, and set up, petitions on issues which are important to you

Environmental Groups

1. Join an active environmental group to carry out positive action, get to know other like-minded people and for your own well-being
2. Organise your own, including a clear vision, a suitable name and logo, and whether you will focus upon local, national or global action
3. Spread awareness of any group you are involved with, along with their actions
4. Carry out fundraising activities to support your chosen environmental work
5. Find a role with an environmental group which suits you; protesting on the streets is not for everyone
6. Work alongside other environmental groups
7. Carry out relevant actions across the other lists in this chapter

Awareness

1. Stay up to date with current environmental information through a variety of media, literature, documentaries, podcasts, meetings, talks and discussions
2. Ensure you include positive environmental news and celebrate it
3. Strive to make informed decisions
4. Get yourself a strong environmental education
5. Speak to your local school about how they are implementing an environmental curriculum as well as positive actions
6. Speak to local and national media about environmental actions
7. Interview environmentalists and share this through different media
8. Document environmental action through photos, video, podcasts, and the written word
9. Listen to true life accounts of people around the world
10. Discuss with others what kind of future you want, and which changes need to occur for this to happen
11. Attend or even organise 'talking books' sessions with environmentalists
12. Involve your friends and family in what you are trying to do

13. Support the learning of others on environmental issues, through respectful conversations, whether in person or online
14. Listen to those who hold different views on Climate Change but also politely put your thoughts across, citing evidence

Energy

1. Increase the insulation in your home, including roof and loft insulation, cavity wall, draught proofing windows and doors, water tank and pipes, floor insulation and double glazing
2. Turn off all appliances when not required to be on, including not leaving them on standby. Even better, unplug them
3. Turn off lights as you leave a room
4. Switch lighting to LEDs where suitable
5. Install Gravitylights or Nowlights
6. Switch to higher efficiency appliances
7. Switch your home electricity to 100% renewables
8. Ensure fridge and freezer coils are clean
9. Switch to activities which require less energy (keep exercising though!)
10. Have a smart energy meter at home
11. Use meters in individual appliances to measure the energy they use

12. Use your own solar and wind where possible, a solar charger is a good start
13. Implement ground source heating or air source heating where possible
14. Introduce and support 'No Electric Days' at work, school and home
15. In rented accommodation, lobby the landlord to make the property more energy efficient
16. Switch to electrical heating and cooking
17. Fit reflecting panels behind the radiators
18. Set thermostat lower in colder times
19. Put on extra layers if cold rather than turn up the heating
20. Reduce use of air conditioning
21. Paint your roof white and/or create a living roof
22. Use an awning in summer to reduce heat going in through windows
23. If needing batteries, use rechargeable

Transport

1. Walk, cycle, or use public transport where possible
2. Organise a carpool if public transport is not available
3. Switch to electric transport rather than fossil fuel based, where possible

4. Switch to hybrid transport if unable to switch to full electric
5. Otherwise purchase cars with higher km per litre
6. Reduce the numbers of cars in the family
7. Ensure tyres are inflated to the correct level
8. Maintain vehicles in good condition, including tyres, oil, and filters
9. Remove excess weight from being stored in the car
10. No idling
11. Accelerate and decelerate smoothly
12. Find out the speed at which your vehicle is most efficient and aim for this, adjusting as suitable for the speed limit
13. Reduce air travel
14. Note that offsetting rarely is as it appears and not flying in the first place is much better
15. In business, choose to video call rather than fly on a plane

Food

1. Reduce consumption of meat, dairy, eggs, and fish
2. Replace beef and lamb for other meats such as pork and chicken
3. Implement meat-free days

4. Introduce vegan/vegetarian meals into your diet
5. Support your local, plant-based farms
6. Support organic food produce
7. Try Veganuary
8. Try non-dairy milks
9. Go vegetarian
10. Go vegan
11. Research different vegan recipes and try them out
12. Attend vegan festivals to both talk to others and try out the food
13. Look at evidence on your own health and plant-based diets, including why more elite athletes have made the switch
14. Speak to where you purchase food and drink about providing more plant-based options
15. Reduce use of all palm oil products and purchase only those which use sustainable palm oil
16. Find out more about the environmental, animal welfare and health differences between different foods
17. Plan your meals and buy food to match
18. Try not to go food shopping hungry
19. Eat all the food that you buy to reduce food waste

20. When 'out and about' take waste-free food with you
21. Share meals
22. Avoid air-freighted food products
23. Eat food in season
24. Avoid foods which have come from recently deforested land
25. Make any significant change in food consumption gradual, rather than all in one go
26. Defrost food in the fridge overnight rather than in the microwave
27. Get creative with leftovers for more delicious meals
28. Eat food up to 'use-by', not 'sell-by' dates
29. Find out which foods can be frozen and do so, as suitable
30. Make a record of food waste and identify improvements
31. Take a reusable container for use as a 'doggy-bag' at restaurants
32. Donate to local food banks
33. Shop at farmers' markets
34. Buy non-perishable foods in bulk to reduce on packaging
35. Repurpose food containers, such as glass, for either storing, or serving food

36. Let fats and oils solidify, then collect and dispose of suitably, rather than down the sink

Wildlife

1. Spend time in nature
2. Get to know your local flora and fauna
3. Consider how grateful you are of the existence of such things are the Sun, the Earth, the atmosphere, your family and friends, trees, flowers, your favourite animal, and others
4. Join a local conservation group
5. Recognise that all species are part of our Earth ecosystem and to protect one is to protect all
6. Join, or even organise, local tree planting
7. Plant native trees and other plants within your local environment
8. Plant native flowers which support pollinators
9. Grow a range of native plants, which flower at different times of the year
10. Rewild your garden
11. Use a watering can rather than a running hose to water plants
12. Know when suitable times of the year are to cut the grass, and use a push lawn mower, or electric if need be

13. Have house plants to reduce pollution
14. Provide suitable bird food, as well as for other local fauna
15. Include a bat box
16. Create a bug hotel
17. Create a wormery
18. Grow your own herbs, fruit, and vegetables
19. If possible, include a garden pond
20. Have hedges where possible, and refrain from trimming them whilst their nuts and berries are available
21. Support, or even organise, local rewilding projects
22. Compost food waste, grass cuttings, and other plant material
23. Avoid peat-based composts
24. Refrain from using chemical pesticides
25. Remove weeds by hand, rather than using chemical weedkillers

Water

1. Consider all water use and find ways to reduce yours
2. Collect rainwater in water butts for use for plants and toilets
3. Use low flush toilets; this could be as simple

as adding a bottle with sand in (not a brick) in the tank

4. Implement low flow faucets on taps and shower heads
5. If you have a dishwasher, use it rather than handwash, but only when full
6. Use a bowl or a plughole when handwashing
7. Wash only full loads of laundry in the washing machine
8. Use water efficient washing machines on the suitable program, including at lower temperatures
9. When boiling a kettle, only put in just enough water
10. When boiling water in a pan, put in just enough, use a lid, and use a setting just high enough for boiling to take place
11. Water plants at night, to reduce evaporation

Bathroom

1. Take showers rather than baths
2. Use a timer to limit shower time
3. Turn off water when not directly using it whilst brushing teeth/washing/shaving
4. Fix leaking taps and pipes
5. Use bar soap, shampoos, and conditioners, rather than gels (which contain microplastics)

6. Research which health, beauty and cleaning products are the best environmentally and strive to use those
7. Use bamboo rather than plastic toothbrushes
8. Use toothpaste from a glass jar rather than a plastic tube
9. Use reusable safety razors
10. If you need a cotton bud, switch from plastic to bamboo
11. Make your own cleaning products from natural materials
12. If it's yellow let it mellow, if it's brown flush it down
13. Switch to toilet paper packaged in paper, not plastic

Material Items

1. Follow the ten Rs; Rethink, Refuse, Reduce, Reuse, Repurpose, Repair, Refurbish, Regift, Recycle and Rot
2. When purchasing new items, choose to support ethical, environmental purchasing
3. Use your own reusable shopping bags
4. No single use bags
5. Avoid items with (excess) packaging
6. When in need of items, such as furniture, buy second-hand or at least sustainably resourced

7. Buy refurbished electronic items, such as phones and computers
8. Determine which recycling can happen in your area and carry it out, as well and letting other others know
9. Ensure items to recycle are empty and clean
10. Squash/flatten items to recycle to save space
11. Take inhalers and unused medicines to a pharmacy for recycling
12. Combine foil into a large ball before putting into the recycling
13. Look after and repair all material items
14. Find out where you can get items repaired locally
15. Share tools and similar with others
16. Get creative and upcycle items
17. Consider why you really want each item you are planning on buying
18. Consider whether you really need a new phone, and similar
19. Aim to only replace items once they can no longer be repaired or have become particularly unfriendly environmentally
20. Sell or donate your old mobile phone and other electronic devices, or find out where you can recycle them locally
21. Pick up litter, especially plastics

22. Organise, or at least take part in litter picking events, including beach clean-ups
23. Learn where you can take hazardous waste, including old paint, oil, and cleaning products
24. Refuse 'give-away' items that you don't want

Plastics

1. Avoid all single use items, including bags, plates, cups, bottles, straws and cutlery, especially those made from plastic
2. Carry a reusable water bottle, as well as other reusable items you may need, including coffee cups, plates, cutlery, straws, etc
3. Have water bottle filling stations set up at work/school
4. Bring reusable mesh bags for fruit and veg
5. Use silicone sealable reusable bags
6. Shop in zero waste stores
7. Strive to purchase items with less plastic packaging
8. Wet wipes, just no
9. Use reusable nappies
10. Use biodegradable refuse sacks
11. Switch from plastic washing up brushes, sponges, and scourers, to those made from coconuts

12. Use a reusable menstrual cup/period pants rather than pads and tampons
13. Keep bubble wrap packaging to use for storing items, insulating plants or sending an item through the post
14. Many places now recycle soft plastics as well as harder ones; find suitable local schemes
15. Switch from packaging of individual portions
16. Snip plastic rings if you have them
17. Only laminate paper if you really want it for hundreds of years
18. Use loose tea rather than teabags, many of which contain plastic
19. Avoid chewing gum and tobacco
20. In you need balloons, fill with air, not helium

Clothing

1. Sell or donate unwanted items, including clothes
2. Attend, or even organise, clothes swaps
3. Use, or even organise, online groups to swap and donate clothes, including school uniforms, shoes, sportswear, etc
4. Buy second-hand clothes
5. Buy higher quality items and make them last
6. Buy clothes made from natural fibres rather than plastics

7. Borrow or rent items rather than buying them
8. Share clothing, and other items, with others
9. Wear environmental logos on badges and clothing
10. Use a clothesline rather than a tumble dryer
11. Use bamboo rather than plastic clothes pegs
12. Use microfibre bags in laundry to catch microfibres from the clothes
13. Use Ecoeggs or detergent strips rather than liquid detergent
14. Use refills for detergents and other products

Paper

1. Read second-hand paper books and use libraries
2. View papers and magazines online rather than buying printed copies
3. Use recycled paper for books, toilet paper, paper towels, etc
4. Contact the Post Office to stop receiving junk mail
5. If you need to wrap presents, use reusable wrapping paper
6. Only print when you really need to
7. Set printers default to double sided

8. Only purchase an e-book reader, rather than paper books, if you read a lot
9. Use reusable cloths rather than paper towels or tissues
10. Use white boards/chalk boards rather than paper for messages

Careers

1. Research green careers and their requirements
2. Talk to others about how their work is linked to the environment
3. Recognise that all jobs can make a positive difference if done in the right way
4. Talk to local companies about what they are doing to support the environment
5. Work with your workplace on their environmental action, including many of the actions elsewhere on this list

Be Creative

1. Express your creativity through environmental writing, poetry, music, drama, and art
2. Take part in, or even organise environmental writing, poetry, music, drama, and art competitions
3. Display creative environmental work

4. Create environmental signs and banners for use in raising awareness and climate protests
5. Write down how the world's environmental issues make you feel and discuss with others
6. Write a letter to your future self, or a future child or grandchild, explaining the world today
7. Write a letter back from your future self, or a future child or grandchild, describing your hopes for what actions are carried out now
8. Make a pledge card and write on three positives changes you will make
9. Share any pledges for positive change that you make

Mindset

1. Be a positive role model yourself, in your thoughts, speech and action
2. Have the environment in mind as you take decisions through the day
3. Switch from 'them and us' to 'all of us'
4. Remember that the more we work together, the more positive change we can achieve
5. Know that all actions add up and can bring positive change
6. Don't judge or bring shame upon others who aren't carrying out the same actions as you
7. Understand that you don't need to know

everything to join an environmental group, or otherwise get involved

8. Don't think that you need a perfect life, you have been brought up in a society in which that was extremely difficult to do
9. Forgive yourself, breaking old habits takes time
10. Connect with people all over the world and listen to their stories
11. Show empathy for others; we haven't lived their lives
12. Look into the reasons why some others may react differently to our current environmental issues and work with them, not against them
13. Share your hope for a better future with those around you
14. Don't get disheartened by deniers. Some people may never come around, or will take a very long time to do so
15. If feeling helpless, remember that whilst you may not be able to do everything, you can do something; every positive change makes a difference
16. Keep in mind that we already know how to make the necessary positives changes, all we need is the political will and collective action
17. When in doubt, look at the evidence, look at the science

18. Take time away from environmentalism on occasion, we all need a break
19. Keep in mind that history is full of examples of positive change brought about by people
20. Remember that it's not that you are protecting the planet, you are part of the planet
21. Recognise that whilst some environmental actions may require sacrifice, if there is no such action then far greater sacrifices will be forced upon us
22. Remember, this is our climate, our future, and there is a better way.

ABOUT THE AUTHOR

Jonny Friend is a UN Accredited Climate Change teacher and Head of Science at a secondary school in Wiltshire, UK. After graduating with a degree in Astrophysics from Cardiff University in 1992, he qualified as a science teacher from Swansea University in 1993. He has also taught in Lincolnshire, Hungary, the Bahamas, Costa Rica, and Somerset.

Jonny has been interested in environmental issues for as long as he can remember. With an increasing need in schools, he launched a successful school environmental group, Schools for Environmental Action (S.E.A), which has raised awareness and been a place of inspiration, hope, support, and action for many young people. Jonny believes firmly in the power of positive change and active hope. Together with staff and students, S.E.A. has inspired decisive action both within and outside of the school.

Jonny lives in Wiltshire with his wife and two sons.

For more information:

www.ourclimateourfuture.co.uk
Twitter: @jonnyfriend1

BIBLIOGRAPHY

Introduction

1. Democracy Now!. 2022. *Greta Thunberg: "Change is Coming, Whether You Like It or Not"*. [online] Available at: <https://www.democracynow.org/2019/9/19/greta_thunberg_change_is_coming_whether> [Accessed 26 January 2022].

Chapter 1: What is this Climate Change anyway?

1. Mann, M., Zhang, Z., Hughes, M., Bradley, R., Miller, S., Rutherford, S. and Ni, F., 2008. Proxy-based reconstructions of hemispheric and global surface temperature variations over the past two millennia. Proceedings of the National Academy of Sciences, 105(36), pp.13252-13257.

2. Bloomberg.com. 2021. *Bloomberg - Are you a robot?*. [online] Available at: <https://www.bloomberg.com/news/articles/2020-12-02/global-temperatures-already-1-2-c-above-pre-industrial-levels> [Accessed 24 October 2021].

3. Climatecentral.org. 2021. *Top 10 Warmest Years on Record*. [online] Available at: <https://www.climatecentral.org/gallery/graphics/top-10-warmest-years-on-record> [Accessed 24 October 2021].

4. Earth Temperature without GHGs [Online]. Energy Education. Available at: energyeducation.ca/encyclopedia/Earth_Temperature_withoutGHGs (Accessed: 24 October 2021).

5. "Understanding Global Warming Potentials | US EPA." N.p, n.d. Web. 24 Oct. 2021

https://www.epa.epa.gov/ghgemissions/understanding-global-warming-potentials>.

6. Gml.noaa.gov. 2021. *Global Monitoring Laboratory - Carbon Cycle Greenhouse Gases*. [online] Available at: <https://gml.noaa.gov/ccgg/trends/> [Accessed 24 October 2021].

7. Gml.noaa.gov. 2021. *Global Monitoring Laboratory - Carbon Cycle Greenhouse Gases*. [online] Available at: <https://gml.noaa.gov/ccgg/trends_ch4/> [Accessed 24 October 2021].

8. Gml.noaa.gov. 2021. *Global Monitoring Laboratory - Carbon Cycle Greenhouse Gases*. [online] Available at: <https://gml.noaa.gov/ccgg/trends_n2o/> [Accessed 24 October 2021].

9. Noaa.gov. 2021. *The changing Arctic: A greener, warmer and increasingly accessible region | National Oceanic and Atmospheric Administration*. [online] Available at: <https://www.noaa.gov/explainers/changing-arctic-greener-warmer-and-increasingly-accessible-region> [Accessed 30 October 2021].

10. Arcticwwf.org. 2021. *Arctic Climate Change*. [online] Available at: <https://arcticwwf.org/work/climate/> [Accessed 10 November 2021].

11. Noaa.gov. 2021. *It's official: July was Earth's hottest month on record | National Oceanic and Atmospheric Administration*. [online] Available at: <https://www.noaa.gov/news/its-official-july-2021-was-earths-hottest-month-on-record> [Accessed 10 November 2021].

12. Reading.ac.uk. 2022. *Climate stripes - University of Reading*. [online] Available at: <https://www.reading.ac.uk/planet/climate-resources/climate-stripes> [Accessed 27 January 2022].

13. Met Office. 2021. *The heatwave of 2003*. [online] Available at: <https://www.metoffice.gov.uk/weather/learn-about/weather/case-studies/heatwave> [Accessed 10 November 2021].

14. The Verge. 2021. *June heatwave was the 'most extreme' on record for North America*. [online] Available at: <https://www.theverge.com/2021/7/30/22602208/june-heatwave-most-extreme-summer-record-north-america> [Accessed 10 November 2021].

15. Ryan, 2021. *2021 heat wave is now the deadliest weather-related event in Washington history.* [online] Kuow.org. Available at: <https://www.kuow.org/stories/heat-wave-death-toll-in-washington-state-jumps-to-112-people> [Accessed 10 November 2021].

16. (UGA), T., (UKSH), U., (EPFL), S. and (EPFL), S., 2021. *Climate change made deadly heatwave 150 times more likely.* [online] Nature.com. Available at: <https://www.nature.com/articles/d41586-021-01869-0> [Accessed 11 November 2021].

17. Internet Geography. 2021. *Storm Christoph.* [online] Available at: <https://www.internetgeography.net/storm-christoph/> [Accessed 11 November 2021].

18. Yale Climate Connections. 2021. *Central Europe staggers toward recovery from catastrophic flooding: more than 200 killed » Yale Climate Connections.* [online] Available at: <https://yaleclimateconnections.org/2021/07/central-europe-staggers-toward-recovery-from-catastrophic-flooding-more-than-200-killed/> [Accessed 12 November 2021].

19. Nypost.com. 2021. *Fresh flooding in central China kills another 21 people.* [online] Available at: <https://nypost.com/2021/08/13/fresh-flooding-in-central-china-kills-another-21-people/> [Accessed 12 November 2021].

20. Phys.org. 2021. *Nigeria has a flooding challenge: Here's why and what can be done.* [online] Available at: <https://phys.org/news/2021-11-nigeria.html> [Accessed 12 November 2021].

21. Jessie Yeung and Esha Mitra, C., 2021. *Floods and landslides kill at least 150 in India and Nepal.* [online] CNN. Available at: <https://edition.cnn.com/2021/10/20/india/india-uttarakhand-kerala-flood-intl-hnk/index.html> [Accessed 12 November 2021].

22. The Tico Times | Costa Rica News | Travel | Real Estate. 2021. *Costa Rica begins evaluating flooding damage.* [online] Available at: <https://ticotimes.net/2021/07/29/costa-rica-begins-evaluating-flooding-damage> [Accessed 12 November 2021].

23. ReliefWeb. 2021. *Pakistan - Flash floods and landslides (PDMA, PMD) (ECHO Daily Flash of 5 October 2021) - Pakistan.* [online] Available at: <https://reliefweb.int/report/pakistan/pakistan-flash-

floods-and-landslides-pdma-pmd-echo-daily-flash-5-october-2021> [Accessed 12 November 2021].

24. Perry, N., 2021. *Hundreds evacuated from catastrophic NZ floods | The New Daily Hundreds evacuated in New Zealand floods*. [online] The New Daily. Available at: <https://thenewdaily.com.au/news/2021/05/31/new-zealand-floods/> [Accessed 12 November 2021].

25. Disasterassist.gov.au. 2021. *Australian disasters*. [online] Available at: <https://www.disasterassist.gov.au/Pages/australian-disasters.aspx> [Accessed 12 November 2021].

26. Nypost.com. 2021. *NYC streets, subways flooded by Ida remnants*. [online] Available at: <https://nypost.com/2021/09/01/nyc-streets-subway-stations-overrun-by-flash-floods/> [Accessed 12 November 2021].

27. the Guardian. 2021. *London flooding: vehicles stranded and tube stations submerged after thunderstorms – video*. [online] Available at: <https://www.theguardian.com/uk-news/video/2021/jul/25/london-flooding-vehicles-stranded-tube-stations-submerged-thunderstorms-video> [Accessed 12 November 2021].

28. ReliefWeb. 2021. *West and Central Africa: Flooding Situation (As of 29 July 2021) - Democratic Republic of the Congo*. [online] Available at: <https://reliefweb.int/report/democratic-republic-congo/west-and-central-africa-flooding-situation-29-july-2021> [Accessed 12 November 2021].

29. Grida.no. 2021. *Number of flood events by continent and decade since 1950 | GRID-Arendal*. [online] Available at: <https://www.grida.no/resources/6062> [Accessed 12 November 2021].

30. Science. 2021. *Climate Change Helped Spark Syrian War, Study Says*. [online] Available at: <https://www.nationalgeographic.com/science/article/150302-syria-war-climate-change-drought> [Accessed 12 November 2021].

31. Ifrc.org. 2021. *Kenya: Drought hunger crisis 2021 | IFRC*. [online] Available at: <https://www.ifrc.org/emergency/kenya-drought-hunger-crisis-2021> [Accessed 12 November 2021].

32. Ifrc.org. 2021. *Kenya: Drought hunger crisis 2021 | IFRC.* [online] Available at: <https://www.ifrc.org/emergency/kenya-drought-hunger-crisis-2021> [Accessed 12 November 2021].

33. Earthobservatory.nasa.gov. 2021. *Drought Conditions Continue in Spring 2021.* [online] Available at: <https://earthobservatory.nasa.gov/images/148123/drought-conditions-continue-in-spring-2021> [Accessed 12 November 2021].

34. Earthobservatory.nasa.gov. 2021. *Brazil Battered by Drought.* [online] Available at: <https://earthobservatory.nasa.gov/images/148468/brazil-battered-by-drought> [Accessed 12 November 2021].

35. Amnesty International. 2021. *Angola: Millions facing hunger, as thousands flee their homes as drought ravages the south of Angola.* [online] Available at: <https://www.amnesty.org/en/latest/press-release/2021/07/angola-millions-facing-hunger-as-thousands-flee-their-homes-as-drought-ravages-the-south-of-angola/> [Accessed 12 November 2021].

36. Lo, J., 2021. *Drought and water mismanagement spark deadly protests in Iran.* [online] Climate Home News. Available at: <https://www.climatechangenews.com/2021/07/28/drought-and-water-mismanagement-spark-deadly-protests-in-iran/> [Accessed 12 November 2021].

37. The Independent. 2021. *Madagascar famine becomes first in history to be caused solely by climate crisis.* [online] Available at: <https://www.independent.co.uk/climate-change/madagascar-famine-climate-crisis-b1888058.html?s=09> [Accessed 12 November 2021].

38. Hughes, L., Dean, A., Steffen, W., Rice, M. and Mullins, G., 2021. *Summer of Crisis | Report | Climate Council.* [online] Climate Council. Available at: <https://www.climatecouncil.org.au/resources/summer-of-crisis/> [Accessed 12 November 2021].

39. Reuters. 2021. *Italian wildfires rage on after 49 degree heat record.* [online] Available at: <https://www.reuters.com/world/europe/italian-wildfires-rage-after-49-degree-heat-record-2021-08-12/> [Accessed 12 November 2021].

40. Cookman, L., 2021. *'We lost everything': Despair and questions as Turkey burns.* [online] Aljazeera.com. Available at:

<https://www.aljazeera.com/news/2021/8/3/turkey-wildfires-despair-and-questions-as-southern-coast-burns> [Accessed 12 November 2021].

41. inews.co.uk. 2021. *How tourist family escaped terrifying Greece wildfires as 'ash rained from the sky'*. [online] Available at: <https://inews.co.uk/news/world/greece-fires-evia-wildfires-tourists-blaze-bomb-ash-1145925> [Accessed 12 November 2021].

42. Newsweek. 2021. *Video of escape from wildfire aboard ferry viewed 1.6M times, likened to disaster movie*. [online] Available at: <https://www.newsweek.com/video-escape-wildfire-aboard-ferry-viewed-16m-times-likened-disaster-movie-1617203> [Accessed 12 November 2021].

43. Fire.ca.gov. 2021. *2021 Fire Season | Welcome to CAL FIRE*. [online] Available at: <https://www.fire.ca.gov/incidents/2021/> [Accessed 12 November 2021].

44. the Guardian. 2022. *US wildfires have killed nearly 20% of world's giant sequoias in two years*. [online] Available at: <https://www.theguardian.com/us-news/2021/nov/19/giant-sequoias-wildfires-killed> [Accessed 19 January 2022].

45. News.stanford.edu. 2021. *Stanford researcher reveals influence of global warming on extreme weather events has been frequently underestimated*. [online] Available at: <https://news.stanford.edu/press-releases/2020/03/18/climate-change-meather-predicted/> [Accessed 12 November 2021].

46. Bloomberg.com. 2021. *Bloomberg - Are you a robot?*. [online] Available at: <https://www.bloomberg.com/news/features/2021-07-22/how-climate-change-impacts-the-jet-stream-and-your-weather> [Accessed 12 November 2021].

47. Gfdl.noaa.gov. 2021. *Historical Atlantic Hurricane and Tropical Storm Records – Geophysical Fluid Dynamics Laboratory*. [online] Available at: <https://www.gfdl.noaa.gov/historical-atlantic-hurricane-and-tropical-storm-records/> [Accessed 12 November 2021].

48. Carbon Brief. 2021. *Global warming has 'changed' spread of tropical cyclones around the world*. [online] Available at: <https://www.carbonbrief.org/global-warming-has-changed-spread-

of-tropical-cyclones-around-the-world> [Accessed 12 November 2021].

49. Holland, G. and Bruyère, C., 2021. *Recent intense hurricane response to global climate change.*

50. Weather.gov. 2021. *Hurricane Katrina - August 2005.* [online] Available at: <https://www.weather.gov/mob/katrina> [Accessed 12 November 2021].

51. Lanington, G., 2021. *Tropical Cyclone Ana: Fiji Suffers Second Deadly Storm in Month | Fiji - Digis Mak.* [online] Digis Mak. Available at: <https://digismak.com/tropical-cyclone-ana-fiji-suffers-second-deadly-storm-in-month-fiji/> [Accessed 12 November 2021].

52. Rand.org. 2021. *Hurricanes Irma and Maria: Impact and Aftermath.* [online] Available at: <https://www.rand.org/hsrd/hsoac/projects/puerto-rico-recovery/hurricanes-irma-and-maria.html> [Accessed 12 November 2021].

53. Sky News. 2021. *Tropical Cyclone Seroja hits Indonesia and East Timor: At least 76 people killed and thousands displaced by flash floods and landslides.* [online] Available at: <https://news.sky.com/story/at-least-76-people-killed-in-floods-and-landslides-as-tropical-cyclone-hits-indonesia-and-east-timor-12266365> [Accessed 12 November 2021].

54. Change, N., 2021. *Arctic Sea Ice Minimum | NASA Global Climate Change.* [online] Climate Change: Vital Signs of the Planet. Available at: <https://climate.nasa.gov/vital-signs/arctic-sea-ice/> [Accessed 12 November 2021].

55. NASA. 2021. *Antarctic Sea Ice Reaches New Record Maximum.* [online] Available at: <https://www.nasa.gov/content/goddard/antarctic-sea-ice-reaches-new-record-maximum> [Accessed 12 November 2021].

56. Nasa.gov. 2021. *NASA - Measuring Gravity With GRACE.* [online] Available at: <https://www.nasa.gov/audience/foreducators/k-4/features/F_Measuring_Gravity_With_Grace.html> [Accessed 12 November 2021].

57. Climate Change: Vital Signs of the Planet. 2021. *Is Antarctica*

melting? – Climate Change: Vital Signs of the Planet. [online] Available at: <https://climate.nasa.gov/news/242/is-antarctica-melting/> [Accessed 12 November 2021].

58. Sky News. 2021. *Almost all world's glaciers losing mass - and at an accelerated pace, report says*. [online] Available at: <https://news.sky.com/story/almost-all-worlds-glaciers-losing-mass-and-at-an-accelerated-pace-report-says-12289660> [Accessed 12 November 2021].

59. King, M., Howat, I., Candela, S., Noh, M., Jeong, S., Noël, B., van den Broeke, M., Wouters, B. and Negrete, A., 2021. *Dynamic ice loss from the Greenland Ice Sheet driven by sustained glacier retreat*.

60. Climate.gov. 2021. *Climate Change: Global Sea Level | NOAA Climate.gov*. [online] Available at: <https://www.climate.gov/news-features/understanding-climate/climate-change-global-sea-level> [Accessed 12 November 2021].

61. the Guardian. 2021. *Five Pacific islands lost to rising seas as climate change hits*. [online] Available at: <https://www.theguardian.com/environment/2016/may/10/five-pacific-islands-lost-rising-seas-climate-change> [Accessed 12 November 2021].

62. IUCN. 2021. *Ocean deoxygenation*. [online] Available at: <https://www.iucn.org/resources/issues-briefs/ocean-deoxygenation> [Accessed 12 November 2021].

63. Frontiers. 2021. *Recent Developments in Oxygen Minimum Zones Biogeochemistry*. [online] Available at: <https://www.frontiersin.org/research-topics/26630/recent-developments-in-oxygen-minimum-zones-biogeochemistry> [Accessed 24 October 2021].

64. Research.noaa.gov. 2021. *Global ocean is absorbing more carbon from emissions - Welcome to NOAA Research*. [online] Available at: <https://research.noaa.gov/article/ArtMID/587/ArticleID/2412> [Accessed 24 October 2021].

65. Global Voices. 2021. *Thousands of flamingos found dead in Central Anatolia*. [online] Available at: <https://globalvoices.org/2021/07/21/thousands-of-flamingos-found-dead-in-central-anatolia/> [Accessed 12 November 2021].

66. the Guardian. 2021. *'Heat dome' probably killed 1bn marine animals on Canada coast, experts say*. [online] Available at: <https://www.theguardian.com/environment/2021/jul/08/heat-dome-canada-pacific-northwest-animal-deaths> [Accessed 12 November 2021].

67. Ritchie, H. and Roser, M., 2021. *Biodiversity*. [online] Our World in Data. Available at: <https://ourworldindata.org/extinctions#are-we-heading-for-a-sixth-mass-extinction> [Accessed 24 October 2021].

68. Ipcc.ch. 2021. *Fifth Assessment Report — IPCC*. [online] Available at: <https://www.ipcc.ch/assessment-report/ar5/> [Accessed 24 October 2021].

69. Ipcc.ch. 2021. *Summary for Policymakers — Global Warming of 1.5 ºC*. [online] Available at: <https://www.ipcc.ch/sr15/chapter/spm/> [Accessed 24 October 2021].

70. Ipcc.ch. 2021. *AR6 Climate Change 2021: The Physical Science Basis — IPCC*. [online] Available at: <https://www.ipcc.ch/report/sixth-assessment-report-working-group-i/> [Accessed 24 October 2021].

Chapter 2: The Power of Denial

1. bbc. 2021. *Trending - The Denial Files - 1. The 'd-words' v the planet - BBC Sounds*. [online] Available at: <https://www.bbc.co.uk/sounds/play/w3ct2yqj> [Accessed 24 October 2021].

2. BBC News. 2021. *'Hide the decline' revisited*. [online] Available at: <https://www.bbc.co.uk/news/science-environment-15538845> [Accessed 24 October 2021].

3. Petitionproject.org. 2021. *Global Warming Petition Project*. [online] Available at: <http://www.petitionproject.org/index.php> [Accessed 10 November 2021].

4. HuffPost. 2021. *The 30,000 Global Warming Petition Is Easily-*

Debunked Propaganda. [online] Available at: <https://www.huffpost.com/entry/the-30000-global-warming_b_243092> [Accessed 10 November 2021].

5. Climate Change: Vital Signs of the Planet. 2021. *Can you explain the urban heat island effect?*. [online] Available at: <https://climate.nasa.gov/faq/44/can-you-explain-the-urban-heat-island-effect/> [Accessed 24 October 2021].

6. Brahic, C., 2021. *Climate myths: The cooling after 1940 shows CO₂ does not cause warming*. [online] New Scientist. Available at: <https://www.newscientist.com/article/dn11639-climate-myths-the-cooling-after-1940-shows-co2-does-not-cause-warming/> [Accessed 24 October 2021].

7. Phys.org. 2021. *What was the Medieval warm period?*. [online] Available at: <https://phys.org/news/2021-04-medieval-period.html> [Accessed 24 October 2021].

8. Nsf.gov. 2021. *Greenland used to be green, and here's why that's bad news*. [online] Available at: <https://www.nsf.gov/news/news_summ.jsp?org=NSF&cntn_id=302508> [Accessed 24 October 2021].

9. NASA. 2021. *Antarctic Sea Ice Reaches New Record Maximum*. [online] Available at: <https://www.nasa.gov/content/goddard/antarctic-sea-ice-reaches-new-record-maximum> [Accessed 24 October 2021].

10. NASA. 2021. *Why NASA Keeps a Close Eye on the Sun's Irradiance*. [online] Available at: <https://www.nasa.gov/topics/solarsystem/features/sun-brightness.html> [Accessed 24 October 2021].

11. Bailey, D., 2021. *There Is No Impending 'Mini Ice Age' – Climate Change: Vital Signs of the Planet*. [online] Climate Change: Vital Signs of the Planet. Available at: <https://climate.nasa.gov/blog/2953/there-is-no-impending-mini-ice-age/> [Accessed 24 October 2021].

12. Mailonline, J.M.F., 2015. Solar activity is not linked to global warming: Sunspot theory of climate change is the result of an ancient error in the data. *Daily Mail Online*. Available at: https://www.dailymail.co.uk/sciencetech/article-3192370/Solar-activity-NOT-linked-global-warming-Sunspot-theory-climate-change-result-ancient-error-data.html [Accessed February 2, 2022].

13. Earthdata.nasa.gov. 2021. *Volcanoes and Climate Change | Earthdata*. [online] Available at: <https://earthdata.nasa.gov/learn/sensing-our-planet/volcanoes-and-climate-change> [Accessed 24 October 2021].

14. Buis, A., 2021. *Why Milankovitch (Orbital) Cycles Can't Explain Earth's Current Warming – Climate Change: Vital Signs of the Planet*. [online] Climate Change: Vital Signs of the Planet. Available at: <https://climate.nasa.gov/blog/2949/why-milankovitch-orbital-cycles-cant-explain-earths-current-warming/> [Accessed 24 October 2021].

15. Battersby, S., 2021. *Earth's wild ride: Our voyage through the Milky Way*. [online] New Scientist. Available at: <https://www.newscientist.com/article/mg21228411-500-earths-wild-ride-our-voyage-through-the-milky-way/> [Accessed 24 October 2021].

16. Buis, A., 2021. *Flip Flop: Why Variations in Earth's Magnetic Field Aren't Causing Today's Climate Change – Climate Change: Vital Signs of the Planet*. [online] Climate Change: Vital Signs of the Planet. Available at: <https://climate.nasa.gov/blog/3104/flip-flop-why-variations-in-earths-magnetic-field-arent-causing-todays-climate-change/> [Accessed 24 October 2021].

17. Climate.gov. 2021. *El Niño and La Niña: Frequently asked questions | NOAA Climate.gov*. [online] Available at: <https://www.climate.gov/news-features/understanding-climate/el-ni%C3%B1o-and-la-ni%C3%B1a-frequently-asked-questions> [Accessed 24 October 2021].

18. Chandler, D., 2021. *Climate myths: It's been far warmer in the past, what's the big deal?*. [online] New Scientist. Available at:

<https://www.newscientist.com/article/dn11647-climate-myths-its-been-far-warmer-in-the-past-whats-the-big-deal/> [Accessed 24 October 2021].

19. Philosophical Transactions of the Royal Society A: Mathematical, Physical and Engineering Sciences. 2021. *How warm was the last interglacial? New model–data comparisons | Philosophical Transactions of the Royal Society A: Mathematical, Physical and Engineering Sciences*. [online] Available at: <https://royalsocietypublishing.org/doi/10.1098/rsta.2013.0097> [Accessed 9 November 2021].

20. Brahic, M., 2021. *Climate myths: Ice cores show CO₂ increases lag behind temperature rises, disproving the link to global warming*. [online] New Scientist. Available at: <https://www.newscientist.com/article/dn11659-climate-myths-ice-cores-show-co2-increases-lag-behind-temperature-rises-disproving-the-link-to-global-warming/> [Accessed 24 October 2021].

21. Solar System Exploration Research Virtual Institute. 2021. *Scientists find possible solution to "faint young Sun paradox"*. [online] Available at: <https://sservi.nasa.gov/articles/scientists-find-possible-solution-to-faint-young-sun-paradox/> [Accessed 24 October 2021].

22. Met Office. 2021. *How is climate linked to extreme weather?*. [online] Available at: <https://www.metoffice.gov.uk/weather/climate/climate-and-extreme-weather> [Accessed 24 October 2021].

23. Environment. 2021. *Sea level rise, facts and information*. [online] Available at: <https://www.nationalgeographic.com/environment/article/sea-level-rise-1> [Accessed 24 October 2021].

24. Dufresne, J. and Saint-Lu, M., 2021. *Positive Feedback in Climate: Stabilization or Runaway, Illustrated by a Simple Experiment.*

25. Powell, A., 2019. Harvard research shows energy use climbs with age and temperature. *Harvard Gazette*. Available at: https://news.harvard.edu/gazette/story/2019/06/harvard-research-shows-energy-use-climbs-with-age-and-temperature/ [Accessed February 2, 2022].

26. Ritchie, H. and Roser, M., 2021. *Biodiversity*. [online] Our World in Data. Available at: <https://ourworldindata.org/mammals> [Accessed 24 October 2021].

Chapter 3: Energy Truths

1. Our World in Data. 2021. *Energy consumption by source*. [online] Available at: <https://ourworldindata.org/grapher/energy-consumption-by-source-and-region?time=1900..2019&country=~OWID_WRL> [Accessed 24 October 2021].

2. Ritchie, H. and Roser, M., 2021. *CO_2 and Greenhouse Gas Emissions*. [online] Our World in Data. Available at: <https://ourworldindata.org/co2-emissions> [Accessed 24 October 2021].

3. Forest Research. 2021. *Carbon emissions of different fuels – Forest Research*. [online] Available at: <https://www.forestresearch.gov.uk/tools-and-resources/fthr/biomass-energy-resources/reference-biomass/facts-figures/carbon-emissions-of-different-fuels/> [Accessed 24 October 2021].

4. US EPA. 2021. *Deepwater Horizon – BP Gulf of Mexico Oil Spill | US EPA*. [online] Available at: <https://www.epa.gov/enforcement/deepwater-horizon-bp-gulf-mexico-oil-spill> [Accessed 24 October 2021].

5. Gerretsen, I., 2021. *Infrared images show under-regulated 'methane crisis' across Europe*. [online] Climate Home News. Available at: <https://www.climatechangenews.com/2021/06/24/infrared-images-

show-regulated-methane-crisis-across-europe/> [Accessed 24 October 2021].

6. Benjamin Storrow, E., 2021. *Methane Leaks Erase Some of the Climate Benefits of Natural Gas*. [online] Scientific American. Available at: <https://www.scientificamerican.com/article/methane-leaks-erase-some-of-the-climate-benefits-of-natural-gas/> [Accessed 24 October 2021].

7. The Independent. 2021. *Lancashire fracking site 'leaked emissions equivalent to 142 transatlantic flights'*. [online] Available at: <https://www.independent.co.uk/climate-change/news/fracking-lancashire-methane-leak-cuadrilla-preston-b1592751.html> [Accessed 24 October 2021].

8. Our World in Data. 2021. *Death rates from energy production per TWh*. [online] Available at: <https://ourworldindata.org/grapher/death-rates-from-energy-production-per-twh> [Accessed 24 October 2021].

9. Bloomberg.com. 2021. *Bloomberg – Are you a robot?*. [online] Available at: <https://www.bloomberg.com/news/features/2019-06-26/a-gas-heist-gone-wrong-an-explosion-and-137-deaths-in-mexico> [Accessed 24 October 2021].

10. UCL News. 2021. *Fossil fuel air pollution responsible for 1 in 5 deaths worldwide*. [online] Available at: <https://www.ucl.ac.uk/news/2021/feb/fossil-fuel-air-pollution-responsible-1-5-deaths-worldwide> [Accessed 24 October 2021].

11. Environmental Health Perspectives. 2021. *Risk Factors for Acute Leukemia in Children: A Review | Environmental Health Perspectives | Vol. 115, No. 1*. [online] Available at: <https://ehp.niehs.nih.gov/doi/10.1289/ehp.9023> [Accessed 24 October 2021].

12. Admin, M., 2021. *When Fossil Fuels Run Out, What Then? – MAHB*. [online] MAHB. Available at: <https://mahb.stanford.edu/library-item/fossil-fuels-run/> [Accessed 24 October 2021].

13. Coady, D., Parry, I., Sears, L. and Shang, B., 2017. How Large Are Global Fossil Fuel Subsidies?. *World Development*, 91, pp.11-27.

14. the Guardian. 2021. *Fossil-fuel industry doubles donations to major parties in four years, report shows.* [online] Available at: <https://www.theguardian.com/environment/2020/feb/12/fossil-fuel-industry-doubles-donations-to-major-parties-in-four-years-report-shows> [Accessed 24 October 2021].

15. World-nuclear.org. 2021. *Radioactive Wastes – Myths and Realities : World Nuclear Association – World Nuclear Association.* [online] Available at: <https://world-nuclear.org/information-library/nuclear-fuel-cycle/nuclear-wastes/radioactive-wastes-myths-and-realities.aspx> [Accessed 24 October 2021].

16. the Guardian. 2021. *Sellafield: 'Everything was contaminated: milk, chickens, the golf course'.* [online] Available at: <https://www.theguardian.com/environment/2012/mar/11/sellafield-stories-book-nuclear-accident> [Accessed 24 October 2021].

17. HISTORY. 2021. *Three Mile Island.* [online] Available at: <https://www.history.com/topics/1970s/three-mile-island> [Accessed 24 October 2021].

18. National Geographic. 2021. *The Chernobyl disaster: what happened, and the long-term impact.* [online] Available at: <https://www.nationalgeographic.co.uk/environment/2019/05/chernobyl-disaster-what-happened-and-long-term-impact> [Accessed 24 October 2021].

19. BBC News. 2021. *Fukushima disaster: What happened at the nuclear plant?.* [online] Available at: <https://www.bbc.co.uk/news/world-asia-56252695> [Accessed 24 October 2021].

20. Marton Dunai, G., 2021. *Nuclear energy too slow, too expensive to save climate: report.* [online] U.S. Available at: <https://www.reuters.com/article/us-energy-nuclearpower-idUSKBN1W909J> [Accessed 24 October 2021].

21. Today, Climate hope as scientists in UK set nuclear fusion record. *TODAY*. Available at: https://www.todayonline.com/world/climate-hope-scientists-uk-set-nuclear-fusion-record-1813801 [Accessed February 10, 2022].

22. Carbon Brief. 2021. *Solar, wind and nuclear have 'amazingly low' carbon footprints, study finds*. [online] Available at: <https://www.carbonbrief.org/solar-wind-nuclear-amazingly-low-carbon-footprints> [Accessed 24 October 2021].

23. the Guardian. 2022. *Geothermal energy could meet a fifth of UK's power needs – report*. [online] Available at: <https://www.theguardian.com/environment/2012/may/30/geothermal-energy-uk-power> [Accessed 23 January 2022].

24. Documents1.worldbank.org. 2021. [online] Available at: <https://documents1.worldbank.org/curated/en/550871468184785413/pdf/106570-ESM-P130625-PUBLIC.pdf> [Accessed 24 October 2021].

25. Nerc.ukri.org. 2022. [online] Available at: <https://nerc.ukri.org/latest/publications/planetearth/planet-earth-2020-issue-1/> [Accessed 23 January 2022].

26. Fast Company. 2021. *Inside The Forgotten Chinese Cities Destroyed By The Three Gorges Dam*. [online] Available at: <https://www.fastcompany.com/3029421/inside-the-forgotten-chinese-cities-destroyed-by-the-three-gorges-dam> [Accessed 24 October 2021].

27. Chatham House – International Affairs Think Tank. 2022. *1. Is Biomass Carbon-neutral?*. [online] Available at: <https://www.chathamhouse.org/2017/02/woody-biomass-power-and-heat/1-biomass-carbon-neutral> [Accessed 25 January 2022].

28. Canadian Centre for Policy Alternatives. 2022. *Suspend approvals of new wood pellet mills in BC, environmental groups and union urge government*. [online] Available at: <https://www.policyalternatives.ca/newsroom/news-releases/suspend-approvals-new-wood-pellet-mills-bc-

environmental-groups-and-union> [Accessed 25 January 2022].

29. Media.voog.com. 2022. [online] Available at: <https://media.voog.com/0000/0037/1265/files/Biomass_report_ENG%20_2020.pdf> [Accessed 25 January 2022].

30. the Guardian. 2022. *Wood burning at home now biggest cause of UK particle pollution*. [online] Available at: <https://www.theguardian.com/environment/2021/feb/16/home-wood-burning-biggest-cause-particle-pollution-fires> [Accessed 25 January 2022].

31. the Guardian. 2022. *Wood burners cause nearly half of urban air pollution cancer risk – study*. [online] Available at: <https://www.theguardian.com/environment/2021/dec/17/wood-burners-urban-air-pollution-cancer-risk-study> [Accessed 25 January 2022].

32. Ecotricity.co.uk. 2022. *Ecotricity explains: Making gas from grass*. [online] Available at: <https://www.ecotricity.co.uk/our-news/2021/ecotricity-explains-making-gas-from-grass> [Accessed 25 January 2022].

Chapter 4: Food, Glorious Food

1. Worldometers.info. 2021. *World Population Clock: 7.9 Billion People (2021) – Worldometer*. [online] Available at: <https://www.worldometers.info/world-population/> [Accessed 25 October 2021].

2. Animalmatters.org. 2021. *Facts – Farm Animals | Animal Matters*. [online] Available at: <https://www.animalmatters.org/facts/farm/> [Accessed 25 October 2021].

3. Ritchie, H. and Roser, M., 2021. *CO_2 and Greenhouse Gas Emissions*. [online] Our World in Data. Available at: <https://ourworldindata.org/emissions-by-sector> [Accessed 25 October 2021].

4. Our World in Data. 2021. *Half of the world's habitable land is used for agriculture*. [online] Available at: <https://ourworldindata.org/global-land-for-agriculture> [Accessed 25 October 2021].

5. Society, N., 2021. *Freshwater Resources*. [online] National Geographic Society. Available at: <https://www.nationalgeographic.org/article/freshwater-resources/> [Accessed 25 October 2021].

6. Ritchie, H. and Roser, M., 2021. *Biodiversity*. [online] Our World in Data. Available at: <https://ourworldindata.org/mammals> [Accessed 25 October 2021].

7. Our World in Data. 2021. *The carbon footprint of foods: are differences explained by the impacts of methane?*. [online] Available at: <https://ourworldindata.org/carbon-footprint-food-methane> [Accessed 25 October 2021].

8. World Wildlife Fund. 2022. *Bittersweet: chocolate's impact on the environment*. [online] Available at: <https://www.worldwildlife.org/magazine/issues/spring-2017/articles/bittersweet-chocolate-s-impact-on-the-environment> [Accessed 25 January 2022].

9. Ritchie, H. and Roser, M., 2021. *Forests and Deforestation*. [online] Our World in Data. Available at: <https://ourworldindata.org/soy> [Accessed 25 October 2021].

10. Farminguk.com. 2021. *Seaweed helps cattle reduce methane emissions by 82%*. [online] Available at: <https://www.farminguk.com/news/seaweed-helps-cattle-reduce-methane-emissions-by-82-_57828.html> [Accessed 25 October 2021].

11. Project Drawdown. 2021. *System of Rice Intensification @ProjectDrawdown #ClimateSolutions*. [online] Available at: <https://www.drawdown.org/solutions/system-of-rice-intensification> [Accessed 25 October 2021].

12. Our World in Data. 2021. *If the world adopted a plant-based diet we would reduce global agricultural land use from 4 to 1 billion hectares.* [online] Available at: <https://ourworldindata.org/land-use-diets> [Accessed 25 October 2021].

13. Ritchie, H. and Roser, M., 2021. *Land Use.* [online] Our World in Data. Available at: <https://ourworldindata.org/land-use> [Accessed 25 October 2021].

14. Carrington, D., 2021. *Livestock provide just 18% of calories but take up 83% of farmland: avoiding meat and dairy is 'single biggest way' to reduce your impact on Earth – The Global Catholic Climate Movement.* [online] The Global Catholic Climate Movement. Available at: <https://catholicclimatemovement.global/livestock-provide-just-18-of-calories-but-take-up-83-of-farmland-avoiding-meat-and-dairy-is-single-biggest-way-to-reduce-your-impact-on-earth/> [Accessed 25 October 2021].

15. World Bank Blogs. 2021. *Chart: Globally, 70% of Freshwater is Used for Agriculture.* [online] Available at: <https://blogs.worldbank.org/opendata/chart-globally-70-freshwater-used-agriculture> [Accessed 25 October 2021].

16. Ritchie, H. and Roser, M., 2021. *Environmental impacts of food production.* [online] Our World in Data. Available at: <https://ourworldindata.org/environmental-impacts-of-food> [Accessed 25 October 2021].

17. Hammed, T.B., Oloruntoba, E.O. & Ana, G.R.E.E., 2019. Enhancing growth and yield of crops with nutrient-enriched organic fertilizer at wet and dry seasons in ensuring climate-smart agriculture – International Journal of recycling of Organic Waste in agriculture. *SpringerLink.* Available at: https://link.springer.com/article/10.1007/s40093-019-0274-6 [Accessed February 6, 2022].

18. the Guardian. 2021. *Humans just 0.01% of all life but have destroyed 83% of wild mammals – study.* [online] Available at: <https://www.theguardian.com/environment/2018/may/21/human-

race-just-001-of-all-life-but-has-destroyed-over-80-of-wild-mammals-study> [Accessed 25 October 2021].

19. Conservation.org. 2021. *Sustainable Palm Oil*. [online] Available at: <https://www.conservation.org/projects/sustainable-palm-oil> [Accessed 25 October 2021].

20. Greenpeace.org.uk. 2021. [online] Available at: <https://www.greenpeace.org.uk/news/5-problems-with-sustainable-palm-oil/> [Accessed 25 October 2021].

21. Project Drawdown. 2021. *Conservation Agriculture @ProjectDrawdown #ClimateSolutions*. [online] Available at: <https://www.drawdown.org/solutions/conservation-agriculture> [Accessed 25 October 2021].

22. Petpedia. 2021. *43 Frightening Bee Statistics & Facts for 2021*. [online] Available at: <https://petpedia.co/bee-statistics/> [Accessed 25 October 2021].

23. Covenant Wildlife. 2021. *Are There Different Types of Honey Bees? – Covenant Wildlife*. [online] Available at: <https://covenantwildlife.com/blog/are-there-different-types-of-honey-bees/> [Accessed 25 October 2021].

24. Shi, Z., Allison, S., He, Y., Levine, P., Hoyt, A., Beem-Miller, J., Zhu, Q., Wieder, W., Trumbore, S. and Randerson, J., 2020. The age distribution of global soil carbon inferred from radiocarbon measurements. *Nature Geoscience*, 13(8), pp.555-559.

25. Humanfuture.net. 2021. [online] Available at: <https://humanfuture.net/sites/default/files/Final%20Report%20on%20Food%20Security_0.pdf> [Accessed 25 October 2021].

26. Nerc.ukri.org. 2022. [online] Available at: <https://nerc.ukri.org/latest/publications/planetearth/planet-earth-2020-issue-1/> [Accessed 23 January 2022].

27. Project Drawdown. 2021. *Nutrient Management @ProjectDrawdown #ClimateSolutions*. [online] Available at:

<https://www.drawdown.org/solutions/nutrient-management> [Accessed 25 October 2021].

28. Sustainableproduction.org. 2021. [online] Available at: <http://sustainableproduction.org/downloads/AgricultureandCancer_001.pdf> [Accessed 25 October 2021].

29. ResearchGate. 2021. *(PDF) Societal Impacts of Desertification: Migration and Environmental Refugees?*. [online] Available at: <https://www.researchgate.net/publication/227120780_Societal_Impacts_of_Desertification_Migration_and_Environmental_Refugees> [Accessed 25 October 2021].

30. Anon, AMTEC: Smart Farming Solutions from Ellis Machinery. *Used Farm Agricultural Machinery for Sale*. Available at: https://amtec-group.com/index.php/blog/pros-and-cons-of-direct-drilling [Accessed January 29, 2022].

31. Rainwaterrunoff.com. 2021. *Project Drawdown – Regenerative Agriculture + Sustainable Food + Rainwater Harvesting*. [online] Available at: <https://rainwaterrunoff.com/tag/project-drawdown/> [Accessed 25 October 2021].

32. Our World in Data. 2021. *You want to reduce the carbon footprint of your food? Focus on what you eat, not whether your food is local.* [online] Available at: <https://ourworldindata.org/food-choice-vs-eating-local> [Accessed 25 October 2021].

33. Our World in Data. 2021. *Very little of global food is transported by air; this greatly reduces the climate benefits of eating local.* [online] Available at: <https://ourworldindata.org/food-transport-by-mode> [Accessed 25 October 2021].

34. American, S., 2021. *Ocean Fish Numbers Cut in Half Since 1970.* [online] Scientific American. Available at: <https://www.scientificamerican.com/article/ocean-fish-numbers-cut-in-half-since-1970/> [Accessed 25 October 2021].

35. Smith, G., 2021. *Overfishing | The Earth Times | Encyclopaedia.* [online] Earthtimes.org. Available at:

<http://www.earthtimes.org/encyclopaedia/environmental-issues/overfishing/> [Accessed 25 October 2021].

36. Environmental Health Perspectives. 2022. *Mercury levels in high-end consumers of fish. | Environmental Health Perspectives | Vol. 111, No. 4.* [online] Available at: <https://ehp.niehs.nih.gov/doi/abs/10.1289/ehp.5837> [Accessed 25 January 2022].

37. Our World in Data. 2021. *Food waste is responsible for 6% of global greenhouse gas emissions.* [online] Available at: <https://ourworldindata.org/food-waste-emissions> [Accessed 25 October 2021].

38. RESET.to. 2021. *Global Food Waste and its Environmental Impact | Green Living.* [online] Available at: <https://en.reset.org/knowledge/global-food-waste-and-its-environmental-impact-09122018> [Accessed 25 October 2021].

39. Alexander, G. and Alexander, G., 2021. *Good, Better, Best: The Climate Impacts of Milks.* [online] Sustainabilityi.org. Available at: <https://sustainabilityi.org/good-better-best-the-climate-impacts-of-milks/> [Accessed 25 October 2021].

40. the Guardian. 2021. *Environmental impact of bottled water 'up to 3,500 times greater than tap water'.* [online] Available at: <https://www.theguardian.com/environment/2021/aug/05/environmental-impact-of-bottled-water-up-to-3500-times-greater-than-tap-water> [Accessed 25 October 2021].

41. Medicalnewstoday.com. 2021. *Zoonosis: Definition and diseases.* [online] Available at: <https://www.medicalnewstoday.com/articles/320618#what-is-zoonosis> [Accessed 25 October 2021].

42. Cassella, C., 2021. *Our Exploitation of Wildlife Is Directly Causing More Viruses to Spread to Humans, Study Shows.* [online] ScienceAlert. Available at: <https://www.sciencealert.com/human-exploitation-has-directly-increased-the-risk-of-animal-diseases-spilling-over-to-humans> [Accessed 25 October 2021].

43. Medicalnewstoday.com. 2021. *The effects of antibiotic use in animals on human health and the drug resistance crisis*. [online] Available at: <https://www.medicalnewstoday.com/articles/323639#Global-use-of-antibiotics-in-animals> [Accessed 25 October 2021].

44. Who.int. 2021. *New report calls for urgent action to avert antimicrobial resistance crisis*. [online] Available at: <https://www.who.int/news/item/29-04-2019-new-report-calls-for-urgent-action-to-avert-antimicrobial-resistance-crisis> [Accessed 25 October 2021].

45. Our World in Data. 2021. *How do we reduce antibiotic resistance from livestock?*. [online] Available at: <https://ourworldindata.org/antibiotic-resistance-from-livestock> [Accessed 25 October 2021].

46. Earth.Org – Past | Present | Future. 2021. *The Cost of Subsidising Agriculture*. [online] Available at: <https://earth.org/agriculture-subsidies/> [Accessed 25 October 2021].

47. Food and Agriculture Organization of the United Nations. 2021. *Hunger*. [online] Available at: <https://www.fao.org/hunger/en/> [Accessed 25 October 2021].

48. Who.int. 2021. *Obesity and overweight*. [online] Available at: <https://www.who.int/news-room/fact-sheets/detail/obesity-and-overweight> [Accessed 25 October 2021].

49. Time. 2022. *Climate, Not Conflict. Madagascar's Famine is the First in Modern History to be Solely Caused by Global Warming*. [online] Available at: <https://time.com/6081919/famine-climate-change-madagascar/> [Accessed 21 January 2022].

50. Harvard Health. 2021. *What's the beef with red meat? – Harvard Health*. [online] Available at: <https://www.health.harvard.edu/staying-healthy/whats-the-beef-with-red-meat> [Accessed 25 October 2021].

Chapter 5: Weapons of Mass Consumption

1. NASA. 2021. *90 Years of Our Changing Views of Earth*. [online] Available at: <https://www.nasa.gov/feature/90-years-of-our-changing-views-of-earth> [Accessed 25 October 2021].

2. Earth Overshoot Day. 2021. *Earth Overshoot Day 2021 Home - #MoveTheDate*. [online] Available at: <https://www.overshootday.org/> [Accessed 25 October 2021].

3. Youtube.com. 2021. [online] Available at: <https://www.youtube.com/watch?v=d2J2qdOrW44&t=69s> [Accessed 25 October 2021].

4. Science History Institute. 2021. *History and Future of Plastics*. [online] Available at: <https://www.sciencehistory.org/the-history-and-future-of-plastics> [Accessed 25 October 2021].

5. Ritchie, H. and Roser, M., 2021. *Plastic Pollution*. [online] Our World in Data. Available at: <https://ourworldindata.org/plastic-pollution> [Accessed 25 October 2021].

6. World Economic Forum. 2021. *This is how long everyday plastic items last in the ocean*. [online] Available at: <https://www.weforum.org/agenda/2018/11/chart-of-the-day-this-is-how-long-everyday-plastic-items-last-in-the-ocean/> [Accessed 25 October 2021].

7. livescience.com. 2021. *Your Plastic Bags Are Releasing Greenhouse Gases*. [online] Available at: <https://www.livescience.com/63274-plastics-release-greenhouse-gases.html> [Accessed 25 October 2021].

8. Biology Letters. 2021. *Plastics everywhere: first evidence of polystyrene fragments inside the common Antarctic collembolan Cryptopygus antarcticus | Biology Letters*. [online] Available at: <https://royalsocietypublishing.org/doi/10.1098/rsbl.2020.0093> [Accessed 25 October 2021].

9. Earth Day. 2021. *Fact Sheet: Plastics in the Ocean - Earth Day*. [online] Available at: <https://www.earthday.org/fact-sheet-plastics-in-the-ocean/> [Accessed 25 October 2021].

10. Sky News. 2021. *A million tonnes of 'lethal' ghost gear left in the seas each year, WWF report claims*. [online] Available at: <https://news.sky.com/story/a-million-tonnes-of-lethal-ghost-gear-left-in-the-seas-each-year-wwf-report-claims-12108850> [Accessed 25 October 2021].

11. Society, N., 2021. *Plastic Bag Found at the Bottom of World's Deepest Ocean Trench*. [online] National Geographic Society. Available at: <https://www.nationalgeographic.org/article/plastic-bag-found-bottom-worlds-deepest-ocean-trench/> [Accessed 25 October 2021].

12. The Ocean Cleanup. 2021. *The Great Pacific Garbage Patch | The Ocean Cleanup*. [online] Available at: <https://theoceancleanup.com/great-pacific-garbage-patch/> [Accessed 25 October 2021].

13. the Guardian. 2022. *Nurdles: the worst toxic waste you've probably never heard of*. [online] Available at: <https://www.theguardian.com/environment/2021/nov/29/nurdles-plastic-pellets-environmental-ocean-spills-toxic-waste-not-classified-hazardous> [Accessed 22 January 2022].

14. Environment. 2021. *This whale had more than 88 pounds of plastic in its stomach when it died*. [online] Available at: <https://www.nationalgeographic.com/environment/article/whale-dies-88-pounds-plastic-philippines> [Accessed 25 October 2021].

15. Unesco.org. 2021. *Facts and figures on marine pollution | United Nations Educational, Scientific and Cultural Organization*. [online] Available at: <http://www.unesco.org/new/en/natural-sciences/ioc-oceans/focus-areas/rio-20-ocean/blueprint-for-the-future-we-want/marine-pollution/facts-and-figures-on-marine-pollution/> [Accessed 25 October 2021].

16. Fast Company. 2021. *Rain and wind bring 1,000 tons of microplastics to U.S. protected lands every year*. [online] Available at: <https://www.fastcompany.com/90515124/rain-and-wind-bring-1000-tons-of-microplastics-to-u-s-protected-lands-every-year> [Accessed 25 October 2021].

17. Global Citizen. 2021. *Bottled Water Has Twice as Many Microplastics as Tap Water: Study*. [online] Available at: <https://www.globalcitizen.org/en/content/water-bottles-have-micro-plastics-in-them/> [Accessed 25 October 2021].

18. Phys.org. 2021. *Do you consume a credit card's worth of plastic every week?*. [online] Available at: <https://phys.org/news/2019-06-consume-credit-card-worth-plastic.html> [Accessed 25 October 2021].

19. Breastcancer.org. 2021. *Exposure to Chemicals in Plastic*. [online] Available at: <https://www.breastcancer.org/risk/factors/plastic> [Accessed 25 October 2021].

20. Huang, J., Wen, B., Xu, L., Ma, H., Li, X., Gao, J. and Chen, Z., 2021. *Micro/nano-plastics cause neurobehavioral toxicity in discus fish (Symphysodon aequifasciatus): Insight from brain-gut-microbiota axis*.

21. Society, N., 2021. *A Whopping 91 Percent of Plastic Isn't Recycled*. [online] National Geographic Society. Available at: <https://www.nationalgeographic.org/article/whopping-91-percent-plastic-isnt-recycled/> [Accessed 25 October 2021].

22. News.yahoo.com. 2021. *Yahoo is now a part of Verizon Media*. [online] Available at: <https://news.yahoo.com/most-plastic-recycling-produces-low-122544762.html> [Accessed 25 October 2021].

23. Latham, K., 2021. *The world's first 'infinite' plastic*. [online] Bbc.com. Available at: <https://www.bbc.com/future/article/20210510-how-to-recycle-any-plastic> [Accessed 25 October 2021].

24. Millerrecycling.com. 2021. [online] Available at: <https://millerrecycling.com/china-sword-policy-2021/> [Accessed 25 October 2021].

25. (www.dw.com), D., 2021. *After China's import ban, where to with the world's waste? | DW | 05.04.2019*. [online] DW.COM. Available at: <https://www.dw.com/en/after-chinas-import-ban-where-to-with-the-worlds-waste/a-48213871> [Accessed 25 October 2021].

26. Bbc.co.uk. 2021. [online] Available at: <https://www.bbc.co.uk/newsround/57154255> [Accessed 25 October 2021].

27. Compassion Australia. 2021. *5 Important Results of the Plastic Bag Ban in Kenya*. [online] Available at: <https://www.compassion.com.au/blog/5-important-results-of-the-plastic-bag-ban-in-kenya> [Accessed 25 October 2021].

28. Varsity Online. 2021. *Think vegware will solve our plastic crisis? Think again.*. [online] Available at: <https://www.varsity.co.uk/science/18123> [Accessed 25 October 2021].

29. Patel, P., Cornwall, W. and DeWeerdt, S., 2021. *New plant-based plastic can be printed—and fully recycled*. [online] Anthropocene | Innovation in the Human Age. Available at: <https://www.anthropocenemagazine.org/2021/02/new-plant-based-plastic-is-also-fully-recyclable/> [Accessed 25 October 2021].

30. Beth Howell Writer @Bethany_Howell_ More from this author Beth has a real passion for green living. She's been absorbed in eco research for over three years, 2021. Top 7 most polluting industries. *The Eco Experts*. Available at: https://www.theecoexperts.co.uk/blog/top-7 most-polluting-industries [Accessed February 8, 2022].

31. Theworldcounts.com. 2021. *The World Counts*. [online] Available at: <https://www.theworldcounts.com/challenges/consumption/clothing/

world-cotton-production-statistics/story> [Accessed 26 October 2021].

32. World Wildlife Fund. 2021. *The Impact of a Cotton T-Shirt | Stories | WWF*. [online] Available at: <https://www.worldwildlife.org/stories/the-impact-of-a-cotton-t-shirt> [Accessed 26 October 2021].

33. Environmental Justice Atlas. 2022. *Cotton production at Aral Sea, Uzbekistan and Turkmenistan | EJAtlas*. [online] Available at: <https://ejatlas.org/conflict/the-aral-sea-dried-due-to> [Accessed 25 January 2022].

34. Ecotricity.co.uk. 2021. *How your outfit impacts your carbon footprint*. [online] Available at: <https://www.ecotricity.co.uk/our-news/2018/the-carbon-footprint-of-getting-dressed> [Accessed 26 October 2021].

35. World Economic Forum. 2021. *These facts show how unsustainable the fashion industry is*. [online] Available at: <https://www.weforum.org/agenda/2020/01/fashion-industry-carbon-unsustainable-environment-pollution/> [Accessed 26 October 2021].

36. Evoke.ie. 2021. *What is fashion 'greenwashing'? The sneaky marketing ploy used by some brands | EVOKE.ie*. [online] Available at: <https://evoke.ie/2020/07/04/style/what-is-fashion-greenwashing-the-sneaky-marketing-ploy-used-by-some-brands> [Accessed 26 October 2021].

37. Greenpeace.org.uk. 2021. [online] Available at: <https://www.greenpeace.org.uk/news/9-reasons-to-quit-fast-fashion-this-black-friday/> [Accessed 26 October 2021].

38. Labs.theguardian.com. 2021. *Child labour in the fashion supply chain*. [online] Available at: <https://labs.theguardian.com/unicef-child-labour/> [Accessed 10 November 2021].

39. Pambazuka.org. 2022. *EAC to continue being US dumping ground for used clothing | Pambazuka News*. [online] Available at:

<https://www.pambazuka.org/resources/eac-continue-being-us-dumping-ground-used-clothing> [Accessed 25 January 2022].

40. Aljazeera.com. 2022. *Photos: Chile's desert dumping ground for fast fashion leftovers*. [online] Available at: <https://www.aljazeera.com/gallery/2021/11/8/chiles-desert-dumping-ground-for-fast-fashion-leftovers> [Accessed 25 January 2022].

41. Unearthed. 2021. *London Fashion Week: Fast facts about fast fashion*. [online] Available at: <https://unearthed.greenpeace.org/2019/09/12/fast-facts-about-fast-fashion/> [Accessed 26 October 2021].

42. Teriin.org. 2021. [online] Available at: <https://www.teriin.org/library/files/Cement-Industry-Trends-Report2017.pdf> [Accessed 26 October 2021].

43. Williams, F. and Williams, F., 2021. *Concrete responsible for 8 per cent of all CO2 emissions, says report*. [online] The Architects' Journal. Available at: <https://www.architectsjournal.co.uk/news/concrete-responsible-for-8-per-cent-of-all-co2-emissions-says-report> [Accessed 26 October 2021].

44. Worldsteel.org. 2021. *worldsteel | Statistics*. [online] Available at: <https://www.worldsteel.org/steel-by-topic/statistics.html> [Accessed 26 October 2021].

45. Steelonthenet.com. 2021. *Steelmaking CO2 carbon dioxide emissions by process step*. [online] Available at: <https://www.steelonthenet.com/kb/co2-emissions.html> [Accessed 26 October 2021].

46. Ritchie, H. and Roser, M., 2021. *CO_2 and Greenhouse Gas Emissions*. [online] Our World in Data. Available at: <https://ourworldindata.org/emissions-by-sector> [Accessed 26 October 2021].

47. Recycle-more.co.uk. 2021. *Recycling | Steel*. [online] Available at: <https://www.recycle-more.co.uk/recycling/steel> [Accessed 26 October 2021].

48. Vetter, D., 2022. *How Sweden Delivered The World's First Fossil Fuel-Free Steel*. [online] Forbes. Available at: <https://www.forbes.com/sites/davidrvetter/2021/08/19/how-sweden-delivered-the-worlds-first-fossil-fuel-free-steel/?sh=5c7577016b55> [Accessed 25 January 2022].

49. Barton, C., Cromarty, H. and Wilson, W., 2021. *Empty housing (England)*. [online] House of Commons Library. Available at: <https://commonslibrary.parliament.uk/research-briefings/sn03012/> [Accessed 26 October 2021].

50. GOV.UK. 2021. *Number of new homes built soars to an 11 year high*. [online] Available at: <https://www.gov.uk/government/news/number-of-new-homes-built-soars-to-an-11-year-high> [Accessed 26 October 2021].

51. The Independent. 2021. *Water torture: 3,300,000,000 litres are lost every single day through*. [online] Available at: <https://www.independent.co.uk/news/uk/home-news/water-torture-3-300-000-000-litres-are-lost-every-single-day-through-leakage-2034999.html> [Accessed 26 October 2021].

52. Ellson, A., 2022. *Bosses make millions at dirty water firms*. [online] Thetimes.co.uk. Available at: <https://www.thetimes.co.uk/article/bosses-make-millions-at-dirty-water-firms-9p0j3kmt0> [Accessed 25 January 2022].

53. NRDC. 2021. *Flint Water Crisis: Everything You Need to Know*. [online] Available at: <https://www.nrdc.org/stories/flint-water-crisis-everything-you-need-know> [Accessed 26 October 2021].

54. EN. 2021. *How many people consume bottled water globally?*. [online] Available at: <https://tappwater.co/en/how-many-people-consume-bottled-water-globally/?amp> [Accessed 26 October 2021].

55. Mercurynews.com. 2021. *Action Line: Almost half of your bottled water comes from the tap – The Mercury News*. [online] Available at: <https://www.mercurynews.com/2011/09/23/action-line-almost-half-of-your-bottled-water-comes-from-the-tap/> [Accessed 26 October 2021].

56. the Guardian. 2021. *Environmental impact of bottled water 'up to 3,500 times greater than tap water'*. [online] Available at: <https://www.theguardian.com/environment/2021/aug/05/environmental-impact-of-bottled-water-up-to-3500-times-greater-than-tap-water> [Accessed 25 October 2021].

57. Theworldcounts.com. 2021. *The World Counts*. [online] Available at: <https://www.theworldcounts.com/challenges/planet-earth/state-of-the-planet/world-waste-facts/story> [Accessed 26 October 2021].

58. Better Meets Reality. 2022. *How Many Times Can You Recycle Different Waste Materials? - Better Meets Reality*. [online] Available at: <https://bettermeetsreality.com/how-many-times-can-you-recycle-different-materials-plastic-paper-metal-glass-etc/> [Accessed 25 January 2022].

Chapter 6: The Holy Grail of Technology

1. UKCCSRC. 2021. *CCS Explained - Carbon Capture - UKCCSRC*. [online] Available at: <https://ukccsrc.ac.uk/ccs-explained/carbon-capture/> [Accessed 26 October 2021].

2. Drax. 2021. *What is direct air carbon capture and storage (DACS)? - Drax*. [online] Available at: <https://www.drax.com/carbon-capture/what-is-direct-air-carbon-capture-and-storage-dacs/> [Accessed 26 October 2021].

3. Main. 2021. *Carbon capture technology has been around for decades—here's why it hasn't taken off*. [online] Available at: <https://energy.mit.edu/news/carbon-capture-technology-has-been-around-for-decades-heres-why-it-hasnt-taken-off/> [Accessed 26 October 2021].

4. Phiddian, E., 2021. *Gorgon's carbon capture shortfall*. [online] Cosmos Magazine. Available at: <https://cosmosmagazine.com/earth/earth-sciences/gorgons-carbon-capture-shortfall/> [Accessed 26 October 2021].

5. Global Witness. 2022. *Hydrogen project lauded by Shell to boost green credentials emits more carbon than a million cars | Global Witness*. [online] Available at: <https://www.globalwitness.org/en/press-releases/shell-plant-emissions-million-cars/> [Accessed 26 January 2022].

6. New Atlas. 2022. *Liquid metal catalyst quickly converts carbon dioxide into solid carbon*. [online] Available at: <https://newatlas.com/environment/liquid-metal-carbon-dioxide-capture-solid/> [Accessed 26 January 2022].

7. Eos. 2021. *Basalts Turn Carbon into Stone for Permanent Storage - Eos*. [online] Available at: <https://eos.org/articles/basalts-turn-carbon-into-stone-for-permanent-storage> [Accessed 27 October 2021].

8. the Guardian. 2021. *Ecotricity founder to grow diamonds 'made entirely from the sky'*. [online] Available at: <https://www.theguardian.com/environment/2020/oct/30/ecotricity-founder-to-grow-diamonds-made-entirely-from-the-sky> [Accessed 27 October 2021].

9. Geoengineering.global. 2021. [online] Available at: <https://geoengineering.global/stratospheric-aerosol-injection/> [Accessed 27 October 2021].

10. Animals. 2021. *Here's What Happened the Day the Dinosaurs Died*. [online] Available at: <https://www.nationalgeographic.com/animals/article/what-happened-day-dinosaurs-died-chicxulub-drilling-asteroid-science> [Accessed 27 October 2021].

11. Chemistry.harvard.edu. 2021. *Addressing uncertainties in solar geoengineering*. [online] Available at:

<https://chemistry.harvard.edu/news/addressing-uncertainties-solar-geoengineering> [Accessed 27 October 2021].

12. Geoengineering.ox.ac.uk. 2021. *Oxford Geoengineering Programme // What is Geoengineering?*. [online] Available at: <http://www.geoengineering.ox.ac.uk/www.geoengineering.ox.ac.uk/what-is-geoengineering/what-is-geoengineering/> [Accessed 27 October 2021].

13. BBC News. 2021. *'Whitest ever' paint reflects 98% of sunlight.* [online] Available at: <https://www.bbc.co.uk/news/science-environment-56749105> [Accessed 27 October 2021].

14. Voelcker, J., Voelcker, J., Contributor, J., Halvorson, B., Edelstein, S., Halvorson, B., Edelstein, S., Halvorson, B., Halvorson, B., Edelstein, S., Halvorson, B., Halvorson, B., Halvorson, B., Edelstein, S. and Edelstein, S., 2021. *1.2 Billion Vehicles On World's Roads Now, 2 Billion By 2035: Report*. [online] Green Car Reports. Available at: <https://www.greencarreports.com/news/1093560_1-2-billion-vehicles-on-worlds-roads-now-2-billion-by-2035-report> [Accessed 27 October 2021].

15. Air-quality.org.uk. 2021. *Vehicle Emissions | Air Pollution | City Diesel | LPG | CNG*. [online] Available at: <http://air-quality.org.uk/26.php> [Accessed 27 October 2021].

16. Katie Hunt and Jen Christensen, C., 2021. *Exposure to polluted air is like smoking a pack a day, study says*. [online] CNN. Available at: <https://edition.cnn.com/2019/08/14/health/air-pollution-emphysema-study-climate-scn/index.html> [Accessed 27 October 2021].

17. Americanidleless.org. 2021. *American Idle Less - "A small turn towards the better."*. [online] Available at: <http://www.americanidleless.org/> [Accessed 27 October 2021].

18. British Lung Foundation. 2021. *How does air pollution affect children's lungs? - British Lung Foundation*. [online] Available at: <https://www.blf.org.uk/support-for-you/risks-to-childrens-lungs/air-pollution> [Accessed 27 October 2021].

19. BASIC thinking International. 2021. *Inductive Charging: Electric Cars With Unlimited Range?*. [online] Available at: <https://www.basicthinking.com/inductive-charging-electric-cars/> [Accessed 27 October 2021].

20. the Guardian. 2021. *Children as young as seven mining cobalt used in smartphones, says Amnesty*. [online] Available at: <https://www.theguardian.com/global-development/2016/jan/19/children-as-young-as-seven-mining-cobalt-for-use-in-smartphones-says-amnesty> [Accessed 27 October 2021].

21. feature, n., (LANL), L., (UH), U., University, Q. and Jude), S., 2021. *Seabed mining is coming — bringing mineral riches and fears of epic extinctions*. [online] Nature.com. Available at: <https://www.nature.com/articles/d41586-019-02242-y> [Accessed 27 October 2021].

22. BERNERS-LEE, M., 2021. *HOW BAD ARE BANANAS?*. [S.l.]: GREYSTONE BOOKS.

23. Our World in Data. 2021. *Which form of transport has the smallest carbon footprint?*. [online] Available at: <https://ourworldindata.org/travel-carbon-footprint> [Accessed 27 October 2021].

24. Yale E360. 2021. *How Airplane Contrails Are Helping Make the Planet Warmer*. [online] Available at: <https://e360.yale.edu/features/how-airplane-contrails-are-helping-make-the-planet-warmer> [Accessed 27 October 2021].

25. Business Traveller. 2021. *Majority of flights taken by a small percentage of flyers – Business Traveller*. [online] Available at: <https://www.businesstraveller.com/business-travel/2021/03/31/majority-of-flights-taken-by-a-small-percentage-of-flyers/> [Accessed 27 October 2021].

26. the Guardian. 2021. *How the billionaire space race could be one giant leap for pollution*. [online] Available at: <https://www.theguardian.com/science/2021/jul/19/billionaires-space-tourism-environment-emissions> [Accessed 27 October 2021].

27. Sourcing Hub. 2021. *Air Freight vs. Sea Freight Carbon Footprint // Environmental Impact of Shipping - Sourcing Hub.* [online] Available at: <https://www.sourcinghub.io/air-freight-vs-sea-freight-carbon-footprint/> [Accessed 27 October 2021].

28. Study Finds. 2021. *Study: Social Media Making People Anti-Social, Jealous*. [online] Available at: <https://www.studyfinds.org/study-finds-social-media-jealousy-facebook/> [Accessed 27 October 2021].

29. Karim, F., Oyewande, A., Abdalla, L., Chaudhry Ehsanullah, R. and Khan, S., 2020. Social Media Use and Its Connection to Mental Health: A Systematic Review. *Cureus*,.

30. Phys.org. 2021. *Emissions from computing and ICT could be worse than previously thought*. [online] Available at: <https://phys.org/news/2021-09-emissions-ict-worse-previously-thought.html> [Accessed 27 October 2021].

31. News.microsoft.com. 2021. *Microsoft by the Numbers: Sustainability*. [online] Available at: <https://news.microsoft.com/sustainability-by-the-numbers/en/carbon-negative> [Accessed 27 October 2021].

32. Sustainability.aboutamazon.com. 2021. [online] Available at: <https://sustainability.aboutamazon.com/carbon-methodology.pdf> [Accessed 27 October 2021].

33. Google Cloud Blog. 2021. *Google achieves four consecutive years of 100% renewable energy | Google Cloud Blog*. [online] Available at: <https://cloud.google.com/blog/topics/sustainability/google-achieves-four-consecutive-years-of-100-percent-renewable-energy> [Accessed 27 October 2021].

34. Trees.ecosia.org. 2021. *Ecosia Trees - A holistic reforestation service for your organization*. [online] Available at: <https://trees.ecosia.org/> [Accessed 27 October 2021].

35. the Guardian. 2021. *Bitcoin's energy usage is huge – we can't afford to ignore it.* [online] Available at: <https://www.theguardian.com/technology/2018/jan/17/bitcoin-electricity-usage-huge-climate-cryptocurrency> [Accessed 28 October 2021].

36. The Conversation. 2021. *Big Pharma emits more greenhouse gases than the automotive industry.* [online] Available at: <https://theconversation.com/big-pharma-emits-more-greenhouse-gases-than-the-automotive-industry-115285> [Accessed 28 October 2021].

37. Npr.org. 2021. *NPR Cookie Consent and Choices.* [online] Available at: <https://www.npr.org/templates/story/story.php?storyId=6082152> [Accessed 28 October 2021].

38. Melhem, Y., 2021. *What is killing India's babies?*. [online] Aljazeera.com. Available at: <https://www.aljazeera.com/features/2016/8/15/what-is-killing-indias-babies> [Accessed 28 October 2021].

39. Historyoflamps.com. 2021. *History of Oil Lamps - Who Invented Oil Lamp?*. [online] Available at: <http://www.historyoflamps.com/lamp-history/history-of-oil-lamps/> [Accessed 28 October 2021].

40. Deciwatt.global. 2021. *GravityLight - light from the lift of a weight.* [online] Available at: <https://deciwatt.global/gravitylight> [Accessed 28 October 2021].

41. Carpenter, S., 2022. *The Race To Develop Plastic-Eating Bacteria.* [online] Forbes. Available at: <https://www.forbes.com/sites/scottcarpenter/2021/03/10/the-race-to-develop-plastic-eating-bacteria/?sh=4e6b9fc27406> [Accessed 26 January 2022].

42. Ozdemir, D., McFadden, C., Bayrakdar, I., Papadopoulos, L. and Bayrakdar, I., 2022. *Could Plastic-Eating Mushrooms Solve Humanity's Plastic Problem?*. [online] Interestingengineering.com.

Available at: <https://interestingengineering.com/could-plastic-eating-mushrooms-solve-humanitys-plastic-problem> [Accessed 26 January 2022].

Chapter 7: Biodiversity; Isn't Nature Wonderful

1. livescience.com. 2021. *What Was the First Life on Earth?*. [online] Available at: <https://www.livescience.com/57942-what-was-first-life-on-earth.html> [Accessed 28 October 2021].

2. Science Advances. 2022. *Accelerated modern human–induced species losses: Entering the sixth mass extinction*. [online] Available at: <https://www.science.org/doi/full/10.1126/sciadv.1400253> [Accessed 10 February 2022].

3. ThoughtCo. 2021. *The 5 (or 6) Major Mass Extinctions*. [online] Available at: <https://www.thoughtco.com/the-5-major-mass-extinctions-4018102> [Accessed 28 October 2021].

4. Worldanimalfoundation.com. 2021. *Extinction Facts by World Animal Foundation*. [online] Available at: <https://www.worldanimalfoundation.com/wild-earth/extinction-crisis/> [Accessed 28 October 2021].

5. Student Materials. 2021. *Unit 1 Reading*. [online] Available at: <https://serc.carleton.edu/integrate/teaching_materials/biosphere/student_materials/unit1_read.html> [Accessed 28 October 2021].

6. Noaa.gov. 2021. *Ocean acidification | National Oceanic and Atmospheric Administration*. [online] Available at: <https://www.noaa.gov/education/resource-collections/ocean-coasts/ocean-acidification> [Accessed 28 October 2021].

7. Yale E360. 2021. *How Long Can Oceans Continue To Absorb Earth's Excess Heat?*. [online] Available at: <https://e360.yale.edu/features/how_long_can_oceans_continue_to_absorb_earths_excess_heat> [Accessed 28 October 2021].

8. US EPA. 2022. *Basic Information about Coral Reefs | US EPA*. [online] Available at: <https://www.epa.gov/coral-reefs/basic-information-about-coral-reefs> [Accessed 10 February 2022].

9. Administrator, C., 2021. *Bleaching Background.* [online] Coral.noaa.gov. Available at: <https://www.coral.noaa.gov/education/bleaching-background.html> [Accessed 28 October 2021].

10. Earthobservatory.nasa.gov. 2022. *Mapping the Decline of Coral Reefs*. [online] Available at: <https://earthobservatory.nasa.gov/features/Coral> [Accessed 10 February 2022].

11. Society, N., 2021. *Great Barrier Reef.* [online] National Geographic Society. Available at: <https://www.nationalgeographic.org/article/great-barrier-reef/> [Accessed 24 October 2021].

12. Amy Woodyatt, C., 2021. *The Great Barrier Reef has lost half its corals within 3 decades*. [online] CNN. Available at: <https://edition.cnn.com/travel/article/great-barrier-reef-coral-loss-intl-scli-climate-scn/index.html> [Accessed 24 October 2021].

13. Australian Geographic. 2021. *Short-nosed sea snake* - Australian Geographic. [online] Available at: <https://www.australiangeographic.com.au/topics/wildlife/2015/06/short-nosed-sea-snake/> [Accessed 24 October 2021].

14. "*17 Endangered Animals in Coral Reef* - DeepOceanFacts.com." N.p, n.d. Web. 24 Oct. 2021 https://deepoceanfacts.com/endangered-animals-in-coral-reef.

15. NOAA. 2021. *Restoring Coral Reefs*. [online] Available at: <https://www.fisheries.noaa.gov/national/habitat-conservation/restoring-coral-reefs> [Accessed 24 October 2021].

16. Iucn.org. 2021. [online] Available at: <https://www.iucn.org/sites/dev/files/content/documents/mangroveloss-brief-4pp-19.10.low_.pdf> [Accessed 24 October 2021].

17. UNDP. 2021. *Mangroves in Thailand are constantly threatened. | UNDP in Thailand.* [online] Available at: <https://www.th.undp.org/content/thailand/en/home/presscenter/pressreleases/2018/08/mangroves-in-thailand-are-constantly-threatened-.html> [Accessed 25 October 2021].

18. Elk Grove Divorce Attorney. 2021. *why are mangroves being destroyed.* [online] Available at: <https://anson-law.com/t2vv4jp5/why-are-mangroves-being-destroyed-806721> [Accessed 24 October 2021].

19. Mangroveactionproject.org. 2021. [online] Available at: <https://mangroveactionproject.org/wp-content/uploads/2018/10/Endangered-Species-PDF-02.05.13-18.24.pdf> [Accessed 24 October 2021].

20. *"SEASPIRACY | Netflix | OFFICIAL WEBSITE." N.p, n.d. Web. 24 Oct.2021 <https:www.seaspiracy.org>.*

21. Medicalnewstoday.com. 2021. *Plastics in seafood: Microplastics, pollution, safety, and more.* [online] Available at: <https://www.medicalnewstoday.com/articles/plastics-in-seafood> [Accessed 24 October 2021].

22. American, S., 2021. *Ocean Fish Numbers Cut in Half Since 1970.* [online] Scientific American. Available at: <https://www.scientificamerican.com/article/ocean-fish-numbers-cut-in-half-since-1970/> [Accessed 24 October 2021].

23. "*Continued decline of an Atlantic cod population: how important is* .. "N.p, n.d. Web. 24 Oct. 2021 https://pubmed.ncbi.nlm.nih.gov/17205904.

24. Ritchie, H. and Roser, M., 2021. *Biodiversity.* [online] Our World in Data. Available at: <https://ourworldindata.org/fish-and-overfishing> [Accessed 24 October 2021].

25. Zoological Society of London (ZSL). 2021. *Crisis in global oceans as populations of marine species halve since 1970.* [online] Available at: <https://www.zsl.org/science/news/crisis-in-global-

oceans-as-populations-of-marine-species-halve-since-1970> [Accessed 28 October 2021].

26. World Wildlife Fund. 2021. *Infographic: Sea Turtles*. [online] Available at: <https://www.worldwildlife.org/pages/infographic-sea-turtles> [Accessed 28 October 2021].

27. Globalslaveryindex.org. 2021. *Fishing | Global Slavery Index*. [online] Available at: <https://www.globalslaveryindex.org/2018/findings/importing-risk/fishing/> [Accessed 28 October 2021].

28. Bbc.co.uk. 2021. [online] Available at: <https://www.bbc.co.uk/newsround/50057552> [Accessed 28 October 2021].

29. Iwc.int. 2021. *Commercial Whaling*. [online] Available at: <https://iwc.int/commercial> [Accessed 28 October 2021].

30. Bluecsolutions.org. 2021. [online] Available at: <http://bluecsolutions.org/dev/wp-content/uploads/WhalesChangingClimate.pdf> [Accessed 28 October 2021].

31. Edition.cnn.com. 2021. *Saving California's kelp forests*. [online] Available at: <https://edition.cnn.com/interactive/2021/07/saving-californias-kelp-forests-cte-spc/> [Accessed 28 October 2021].

32. Beautiful, I., 2021. *Seaweed: Food, Fertilizer, Feed & Fuel — Information is Beautiful*. [online] Information is Beautiful. Available at: <https://informationisbeautiful.net/2019/seaweed-food-fertilizer-feed-fuel/> [Accessed 28 October 2021].

33. WWF. 2021. *Planting hope - How seagrass can tackle climate change*. [online] Available at: <https://www.wwf.org.uk/what-we-do/planting-hope-how-seagrass-can-tackle-climate-change> [Accessed 28 October 2021].

34. Wildlifetrusts.org. 2021. *Seagrass | The Wildlife Trusts*. [online] Available at:

<https://www.wildlifetrusts.org/habitats/marine/seagrass> [Accessed 28 October 2021].

35. Earthobservatory.nasa.gov. 2021. *What are Phytoplankton?*. [online] Available at: <https://earthobservatory.nasa.gov/features/Phytoplankton> [Accessed 28 October 2021].

36. Lauren Morello, C., 2021. *Phytoplankton Population Drops 40 Percent Since 1950.* [online] Scientific American. Available at: <https://www.scientificamerican.com/article/phytoplankton-

37. Campaign, G. and habitats, P., 2021. *Forests: Why are forests so important, and how do we best protect them?*. [online] Fauna-flora.org. Available at: <https://www.fauna-flora.org/environments/forests> [Accessed 28 October 2021].

38. Anon, Timber in the carbon economy. *Timber NSW*. Available at: https://timbernsw.com.au/timber-in-the-carbon-economy/ [Accessed February 13, 2022].

39. World Wildlife Fund. 2021. *Forest Habitat | Habitats | WWF*. [online] Available at: <https://www.worldwildlife.org/habitats/forest-habitat> [Accessed 29 October 2021].

40. Ritchie, H. and Roser, M., 2021. *Forests and Deforestation*. [online] Our World in Data. Available at: <https://ourworldindata.org/forest-area> [Accessed 28 October 2021].

41. Our World in Data. 2021. *The world has lost one-third of its forest, but an end of deforestation is possible*. [online] Available at: <https://ourworldindata.org/world-lost-one-third-forests> [Accessed 28 October 2021].

42. Tigers-world.com. 2021. *Tigers Endangered - Tiger Facts and Information*. [online] Available at: <https://www.tigers-world.com/tigers-endangered/> [Accessed 29 October 2021].

43. Faith Karimi, C., 2021. *There are more tigers in captivity in the US than in the wild.* [online] CNN. Available at: <https://edition.cnn.com/2021/05/13/us/tigers-captive-us-wild-trnd/index.html> [Accessed 29 October 2021].

44. World Wildlife Fund. 2021. *African Forest Elephant | Species | WWF.* [online] Available at: <https://www.worldwildlife.org/species/african-forest-elephant> [Accessed 29 October 2021].

45. Rainforest Concern. 2021. *Why are rainforests being destroyed?.* [online] Available at: <https://www.rainforestconcern.org/forest-facts/why-are-rainforests-being-destroyed> [Accessed 29 October 2021].

46. Ritchie, H. and Roser, M., 2021. *Forests and Deforestation.* [online] Our World in Data. Available at: <https://ourworldindata.org/deforestation> [Accessed 29 October 2021].

47. Mongabay. 2021. *Why are rainforests being destroyed?.* [online] Available at: <https://rainforests.mongabay.com/kids/elementary/501.html> [Accessed 29 October 2021].

48. Forestresearch.gov.uk. 2021. [online] Available at: <https://www.forestresearch.gov.uk/documents/1437/QJF_legacy_of_fragmentation_may06.pdf> [Accessed 29 October 2021].

49. Rainforests.mongabay.com. 2021. *Brazil Forest Information and Data.* [online] Available at: <https://rainforests.mongabay.com/deforestation/2000/Brazil.htm> [Accessed 29 October 2021].

50. Climate-Diplomacy. 2021. *Forty years of tree-planting in China: successes and failures.* [online] Available at: <https://climate-diplomacy.org/magazine/environment/forty-years-tree-planting-china-successes-and-failures> [Accessed 29 October 2021].

51. Great Green Wall. 2021. *Great Green Wall — The Great Green Wall.* [online] Available at: <https://www.greatgreenwall.org/about-great-green-wall> [Accessed 29 October 2021].

52. Society, N., 2021. *Grasslands Explained.* [online] National Geographic Society. Available at: <https://www.nationalgeographic.org/article/grasslands-explained/> [Accessed 29 October 2021].

53. Huge carbon stores under grasslands discovered. 2021. *Huge carbon stores under grasslands discovered.* [online] Available at: <https://www.manchester.ac.uk/discover/news/huge-carbon-stores-under-grasslands-discovered/> [Accessed 29 October 2021].

54. Hungate, B., Barbier, E., Ando, A., Marks, S., Reich, P., van Gestel, N., Tilman, D., Knops, J., Hooper, D., Butterfield, B. and Cardinale, B., 2021. *The economic value of grassland species for carbon storage.*

55. Environment. 2021. *Grasslands threats and solutions, facts and information.* [online] Available at: <https://www.nationalgeographic.com/environment/article/grassland-threats> [Accessed 29 October 2021].

56. euronews. 2021. *In pictures: The guardians of the last two northern white rhinos.* [online] Available at: <https://www.euronews.com/green/2021/10/26/guardians-of-the-northern-white-rhino-the-world-s-most-endangered-species> [Accessed 29 October 2021].

57. Csus.edu. 2021. [online] Available at: <https://www.csus.edu/college/social-sciences-interdisciplinary-studies/environmental-studies/_internal/_documents/an-assessment-of-the-international-status-of-the-burrowing-owl-athene-cunicularia---rose-sanchez.pdf> [Accessed 29 October 2021].

58. Environment. 2021. *The Great Plains prairie needs fire to survive. These ranchers are bringing it back..* [online] Available at: <https://www.nationalgeographic.com/environment/article/great-

plains-prairie-needs-fire-to-survive-these-ranchers-are-bringing-it-back> [Accessed 29 October 2021].

59. Encyclopedia Britannica. 2021. *5 Amazing Adaptations of Pyrophytic Plants*. [online] Available at: <https://www.britannica.com/list/5-amazing-adaptations-of-pyrophytic-plants> [Accessed 29 October 2021].

60. Think Global Green. 2021. *Wetlands*. [online] Available at: <https://thinkglobalgreen.org/wetlands/> [Accessed 30 October 2021].

61. Tropicalforestnetwork.com. 2021. *Biodiversity of Wetlands*. [online] Available at: <http://www.tropicalforestnetwork.com/wetlands.html> [Accessed 29 October 2021].

62. People's Trust for Endangered Species. 2021. *Water voles - People's Trust for Endangered Species*. [online] Available at: <https://ptes.org/campaigns/water-voles/> [Accessed 30 October 2021].

63. The RSPB. 2021. *Beaver Reintroduction in the UK - The RSPB*. [online] Available at: <https://www.rspb.org.uk/our-work/policy-insight/species/beaver-reintroduction-in-the-uk/> [Accessed 30 October 2021].

64. American Rivers. 2021. *Why Restoring Wetlands Makes Good Economic Sense - American Rivers*. [online] Available at: <https://www.americanrivers.org/2016/07/economic-case-restoring-wetlands-plan-2014/> [Accessed 30 October 2021].

65. Landscape News. 2021. *Peatlands store twice as much carbon as all the world's forests - Landscape News*. [online] Available at: <https://news.globallandscapesforum.org/viewpoint/peatlands-store-twice-as-much-carbon-as-all-the-worlds-forests/> [Accessed 30 October 2021].

66. Society, N., 2021. *Tundra Biome*. [online] National Geographic Society. Available at:

<https://www.nationalgeographic.org/encyclopedia/tundra-biome/> [Accessed 30 October 2021].

67. Arctic Program. 2021. *Permafrost and the Global Carbon Cycle*. [online] Available at: <https://arctic.noaa.gov/Report-Card/Report-Card-2019/ArtMID/7916/ArticleID/844/Permafrost-and-the-Global-Carbon-Cycle> [Accessed 30 October 2021].

68. Awi.de. 2021. *Single view - AWI*. [online] Available at: <https://www.awi.de/en/about-us/service/press/single-view/aeltester-permafrostboden-von-sibirien-entdeckt.html> [Accessed 31 October 2021].

69. Nps.gov. 2021. *Undergraduates Measure Methane Released by Thawing Permafrost (U.S. National Park Service)*. [online] Available at: <https://www.nps.gov/articles/denali-methane-from-permafrost.htm> [Accessed 31 October 2021].

70. The World from PRX. 2021. *A bold plan to slow the melt of Arctic permafrost could help reverse global warming*. [online] Available at: <https://theworld.org/stories/2017-04-30/bold-plan-slow-melt-arctic-permafrost-could-help-reverse-global-warming> [Accessed 31 October 2021].

71. Whelan, N., Geography, W. and World, T., 2021. *The 10 Largest Deserts In The World*. [online] WorldAtlas. Available at: <https://www.worldatlas.com/articles/10-largest-deserts-in-the-world.html> [Accessed 31 October 2021].

72. Treehugger. 2021. *17 Animals Amazingly Adapted to Thrive in Deserts*. [online] Available at: <https://www.treehugger.com/animals-amazingly-adapted-to-thrive-in-deserts-4869269> [Accessed 31 October 2021].

73. Earthobservatory.nasa.gov. 2021. *Defining Desertification*. [online] Available at: <https://earthobservatory.nasa.gov/features/Desertification/desertification2.php> [Accessed 31 October 2021].

74. Environment. 2021. *Desertification facts and information.* [online] Available at: <https://www.nationalgeographic.com/environment/article/desertification> [Accessed 31 October 2021].

75. BBC Reel. 2022. *The man-made Greek forest that mirrors the Amazon - BBC Reel.* [online] Available at: <https://www.bbc.com/reel/video/p0b24z7f/the-man-made-greek-forest-that-mirrors-the-amazon> [Accessed 26 January 2022].

76. Ollerton, J., Winfree, R. and Tarrant, S., 2021. *How many flowering plants are pollinated by animals?.*

77. University of Cambridge. 2021. *Pollinators: first global risk index for species declines and effects on humanity.* [online] Available at: <https://www.cam.ac.uk/stories/pollinatorsriskindex> [Accessed 31 October 2021].

78. the Guardian. 2021. *'Bee-killing' pesticide now will not be used on UK sugar beet fields.* [online] Available at: <https://www.theguardian.com/environment/2021/mar/03/bee-killing-pesticide-now-will-not-be-used-on-uk-sugar-beet-fields> [Accessed 31 October 2021].

79. the Guardian. 2021. *Climate change is disrupting flower pollination, research shows.* [online] Available at: <https://www.theguardian.com/environment/2014/nov/06/climate-change-is-disrupting-flower-pollination-research-shows> [Accessed 31 October 2021].

80. Gagné-Acoulon, S., 2021. *UN: Pangolin is the World's Most Trafficked Mammal.* [online] Occrp.org. Available at: <https://www.occrp.org/en/daily/12235-un-pangolin-is-the-world-s-most-trafficked-mammal> [Accessed 31 October 2021].

81. Dr Peta HitchensSenior Research Fellow, U., Professor Christine K JohnsonSchool of Veterinary Medicine, U., Hitchens, D. and Johnson, P., 2021. *Don't blame the pangolin (or any other animal) for COVID-19.* [online] Pursuit. Available at:

<https://pursuit.unimelb.edu.au/articles/don-t-blame-the-pangolin-or-any-other-animal-for-covid-19> [Accessed 31 October 2021].

Chapter 8: People are People

1. Hogg, M., 2021. *Self-uncertainty and group identification: Consequences for social identity, group behavior, intergroup relations, and society.*

2. PsyBlog. 2021. *8 Social Psychology Studies Showing People Are Naturally Good - PsyBlog.* [online] Available at: <https://www.spring.org.uk/2021/05/social-psychology-people-naturally-good.php> [Accessed 14 November 2021].

3. Bwcentral.org. 2021. *STUDY: Narcissists More Likely To Be Politicians... | BWCentral.* [online] Available at: <https://bwcentral.org/2020/09/study-narcissists-more-likely-to-be-politicians/> [Accessed 14 November 2021].

4. Assets.cambridge.org. 2021. [online] Available at: <https://assets.cambridge.org/052193/4117/full_version/0521934117_pub.pdf> [Accessed 5 December 2021].

5. Encyclopedia Britannica. 2021. *Industrial Revolution | Definition, History, Dates, Summary, & Facts.* [online] Available at: <https://www.britannica.com/event/Industrial-Revolution> [Accessed 5 December 2021].

6. Industrial Revolution. 2021. *Factories During Industrial Revolution - Conditions & Growth.* [online] Available at: <https://industrialrevolution.org.uk/factories-industrial-revolution/> [Accessed 5 December 2021].

7. ThoughtCo. 2021. *How the Demand for Coal Impacted the Industrial Revolution.* [online] Available at: <https://www.thoughtco.com/coal-in-the-industrial-revolution-1221634> [Accessed 5 December 2021].

8. The climate of the United States and its endemic influences: based chiefly on the records of the medical department and Adjutant

General's office, U., 2021. *The climate of the United States and its endemic influences: based chiefly on the records of the medical department and Adjutant General's office, United States Army - Digital Collections - National Library of Medicine.* [online] Collections.nlm.nih.gov. Available at: <https://collections.nlm.nih.gov/catalog/nlm:nlmuid-0356062-bk> [Accessed 5 December 2021].

9. Climate.gov. 2021. *Happy 200th birthday to Eunice Foote, hidden climate science pioneer | NOAA Climate.gov.* [online] Available at: <https://www.climate.gov/news-features/features/happy-200th-birthday-eunice-foote-hidden-climate-science-pioneer> [Accessed 5 December 2021].

10. Earthobservatory.nasa.gov. 2021. *John Tyndall.* [online] Available at: <https://earthobservatory.nasa.gov/features/Tyndall> [Accessed 5 December 2021].

11. Earthobservatory.nasa.gov. 2021. *Svante Arrhenius.* [online] Available at: <https://earthobservatory.nasa.gov/features/Arrhenius/arrhenius_2.php> [Accessed 5 December 2021].

12. Asme.org. 2021. *The Engineer Who Discovered Global Warming - ASME.* [online] Available at: <https://www.asme.org/topics-resources/content/the-engineer-who-discovered-global-warming> [Accessed 5 December 2021].

13. Heimann, M., 2021. *Charles David Keeling 1928–2005.*

14. Hotez, P., 2021. *The Antiscience Movement Is Escalating, Going Global and Killing Thousands.* [online] Scientific American. Available at: <https://www.scientificamerican.com/article/the-antiscience-movement-is-escalating-going-global-and-killing-thousands/> [Accessed 5 December 2021].

15. The Conversation. 2021. *The '97% climate consensus' is over. Now it's well above 99% (and the evidence is even stronger than that).* [online] Available at: <https://theconversation.com/the-97-

climate-consensus-is-over-now-its-well-above-99-and-the-evidence-is-even-stronger-than-that-170370> [Accessed 5 December 2021].

16. Yamamoto, T. and Wilson, W., n.d. *Hagakure*.

17. Yang, Z., Luo, X. and Jia, H., 2021. *Is It All a Conspiracy? Conspiracy Theories and People's Attitude to COVID-19 Vaccination*.

18. Bbc.com. 2021. *BBC Board*. [online] Available at: <https://www.bbc.com/aboutthebbc/whoweare/bbcboard> [Accessed 5 December 2021].

19. Times News UK. 2021. *Culture Secretary Nadine Dorries' fury at Nick Robinson after he told PM to 'stop talking' - Times News UK*. [online] Available at: <https://timesnewsuk.com/uk-news/culture-secretary-nadine-dorries-fury-at-nick-robinson-after-he-told-pm-to-stop-talking/> [Accessed 5 December 2021].

20. Pedestrian TV. 2021. *Here's Exactly How Far Rupert Murdoch's Media Empire Stretches In Australia And Beyond*. [online] Available at: <https://www.pedestrian.tv/news/rupert-murdoch-news-corp-control/> [Accessed 5 December 2021].

21. Psychology Today. 2021. *Who Controls the Media? You'd Be Surprised*. [online] Available at: <https://www.psychologytoday.com/us/blog/all-is-well/202104/who-controls-the-media-you-d-be-surprised> [Accessed 5 December 2021].

22. Google Books. 2021. *The SAGE Handbook of Sociology*. [online] Available at: <https://books.google.com/books/about/The_SAGE_Handbook_of_Sociology.html?id=pfP4twfRoK8C> [Accessed 5 December 2021].

23. Medium. 2021. *What Is Media Framing and Why Is It So Effective?*. [online] Available at: <https://medium.com/share-the-wealth/what-is-media-framing-and-why-is-it-so-effective-994ad8367381> [Accessed 5 December 2021].

24. Google Books. 2021. *Why We Disagree about Climate Change*. [online] Available at: <https://books.google.co.uk/books?id=DWMkb4ondbQC> [Accessed 5 December 2021].

25. Climatechangecommunication.org. 2021. *America Misled: How the fossil fuel industry deliberately misled Americans about climate change | Center For Climate Change Communication*. [online] Available at: <https://www.climatechangecommunication.org/america-misled/381> [Accessed 19 December 2021].

26. Dictionary.cambridge.org. 2021. *greenwashing*. [online] Available at: <https://dictionary.cambridge.org/dictionary/english/greenwashing> [Accessed 19 December 2021].

27. the Guardian. 2021. *The Volkswagen emissions scandal explained*. [online] Available at: <https://www.theguardian.com/business/ng-interactive/2015/sep/23/volkswagen-emissions-scandal-explained-diesel-cars> [Accessed 19 December 2021].

28. the Guardian. 2021. *The dark secrets behind big oil's climate pledges*. [online] Available at: <https://www.theguardian.com/environment/2021/nov/04/dark-secrets-big-oil-climate-pledges-greenwashing> [Accessed 19 December 2021].

29. Barrett, A. and Barrett, V., 2021. *NGO Sues Coca Cola for Greenwashing (FREE)*. [online] Bioplastics News. Available at: <https://bioplasticsnews.com/2021/06/12/ngo-sues-coca-cola-greenwashing/> [Accessed 19 December 2021].

30. Skinner, C., 2021. *Greenwashing the system: which banks are worst? - Chris Skinner's blog*. [online] Chris Skinner's blog. Available at: <https://thefinanser.com/2021/05/greenwashing-the-system-which-banks-are-worst.html/> [Accessed 19 December 2021].

31. the Guardian. 2021. *Fossil fuel companies paying top law firms millions to 'dodge responsibility'*. [online] Available at: <https://www.theguardian.com/environment/2021/oct/09/fossil-fuel-companies-law-firms> [Accessed 19 December 2021].

32. the Guardian. 2021. *How the BBC let climate deniers walk all over it | George Monbiot*. [online] Available at: <https://www.theguardian.com/commentisfree/2021/jul/08/bbc-climate-change-deniers-fossil-fuel-broadcasters> [Accessed 19 December 2021].

33. Abarbanel, A., 2021. *Is the World Getting Warmer? | The Saturday Evening Post*. [online] The Saturday Evening Post. Available at: <https://www.saturdayeveningpost.com/reprints/warmer-world/> [Accessed 19 December 2021].

34. Paperspast.natlib.govt.nz. 2021. [online] Available at: <https://paperspast.natlib.govt.nz/newspapers/ROTWKG19120814.2.56.5> [Accessed 21 December 2021].

35. the Guardian. 2021. *Fossil-fuel industry doubles donations to major parties in four years, report shows*. [online] Available at: <https://www.theguardian.com/environment/2020/feb/12/fossil-fuel-industry-doubles-donations-to-major-parties-in-four-years-report-shows> [Accessed 27 December 2021].

36. Doering, C., 2021. *Where they stand: Political donations from 10 of the largest food and beverage makers*. [online] Food Dive. Available at: <https://www.fooddive.com/news/where-they-stand-political-donations-from-10-of-the-largest-food-and-bever/585516/> [Accessed 27 December 2021].

37. International Banker. 2021. *Banks and Politics*. [online] Available at: <https://internationalbanker.com/banking/banks-and-politics/> [Accessed 27 December 2021].

38. Dictionary.cambridge.org. 2021. *capitalism*. [online] Available at: <https://dictionary.cambridge.org/dictionary/english/capitalism> [Accessed 27 December 2021].

39. Consulting, D., 2021. *11 Reasons Why Profit is Essential.* [online] Dan Lacy. Available at: <https://dynastybusinessconsulting.com/11-reasons-why-profit-is-essential/> [Accessed 27 December 2021].

40. Vandana and Lenka, U., 2021. *A Review on the Role of Media in Increasing Materialism among Children.*

41. Grantham Research Institute on climate change and the environment. 2021. *The Economics of Climate Change: The Stern Review - Grantham Research Institute on climate change and the environment.* [online] Available at: <https://www.lse.ac.uk/GranthamInstitute/publication/the-economics-of-climate-change-the-stern-review/> [Accessed 28 December 2021].

42. World Economic Forum. 2021. *The $86 trillion world economy – in one chart.* [online] Available at: <https://www.weforum.org/agenda/2019/09/fifteen-countries-represent-three-quarters-total-gdp/> [Accessed 28 December 2021].

43. Data.worldbank.org. 2021. *GDP growth (annual %) | Data.* [online] Available at: <https://data.worldbank.org/indicator/NY.GDP.MKTP.KD.ZG> [Accessed 28 December 2021].

44. the Guardian. 2021. *World's 26 richest people own as much as poorest 50%, says Oxfam.* [online] Available at: <https://www.theguardian.com/business/2019/jan/21/world-26-richest-people-own-as-much-as-poorest-50-per-cent-oxfam-report> [Accessed 28 December 2021].

45. American Chemical Society. 2021. *Chlorofluorocarbons and Ozone Depletion - American Chemical Society.* [online] Available at: <https://www.acs.org/content/acs/en/education/whatischemistry/landmarks/cfcs-ozone.html> [Accessed 28 December 2021].

46. Ozone.unep.org. 2021. *The Montreal Protocol on Substances that Deplete the Ozone Layer | Ozone Secretariat.* [online] Available at:

<https://ozone.unep.org/treaties/montreal-protocol/montreal-protocol-substances-deplete-ozone-layer> [Accessed 28 December 2021].

47. UN News. 2021. *Ozone on track to heal completely in our lifetime, UN environment agency declares on World Day.*. [online] Available at: <https://news.un.org/en/story/2019/09/1046452> [Accessed 28 December 2021].

48. Ipcc.ch. 2021. *IPCC — Intergovernmental Panel on Climate Change.* [online] Available at: <https://www.ipcc.ch/> [Accessed 28 December 2021].

49. Ipcc.ch. 2021. *Reports — IPCC.* [online] Available at: <https://www.ipcc.ch/reports/> [Accessed 28 December 2021].

50. Ipcc.ch. 2021. [online] Available at: <https://www.ipcc.ch/site/assets/uploads/2021/07/AR6_FS_select.pdf > [Accessed 28 December 2021].

51. Ipcc.ch. 2021. *Climate change widespread, rapid, and intensifying – IPCC — IPCC.* [online] Available at: <https://www.ipcc.ch/2021/08/09/ar6-wg1-20210809-pr/> [Accessed 31 December 2021].

52. Unfccc.int. 2021. [online] Available at: <https://unfccc.int/kyoto_protocol> [Accessed 31 December 2021].

53. Unfccc.int. 2021. [online] Available at: <https://unfccc.int/process-and-meetings/the-paris-agreement/the-paris-agreement> [Accessed 31 December 2021].

54. Green Climate Fund. 2021. *Resource Mobilisation.* [online] Available at: <https://www.greenclimate.fund/about/resource-mobilisation> [Accessed 31 December 2021].

55. Ipcc.ch. 2021. *Global Warming of 1.5 ℃ —.* [online] Available at: <https://www.ipcc.ch/sr15/> [Accessed 31 December 2021].

56. Corporateaccountability.org. 2022. [online] Available at: <https://www.corporateaccountability.org/wp-

content/uploads/2021/06/The-Big-Con_EN.pdf> [Accessed 2 January 2022].

57. UN Climate Change Conference (COP26) at the SEC – Glasgow 2021. 2021. *UN Climate Change Conference (COP26) at the SEC – Glasgow 2021*. [online] Available at: <https://ukcop26.org/> [Accessed 31 December 2021].

58. Youtube.com. 2021. [online] Available at: <https://www.youtube.com/watch?v=VMvzJu79WG0&t=13s> [Accessed 31 December 2021].

59. the Guardian. 2022. *Alok Sharma 'deeply frustrated' by India and China over coal*. [online] Available at: <https://www.theguardian.com/environment/2021/nov/14/alok-sharma-deeply-frustrated-by-india-and-china-over-coal> [Accessed 2 January 2022].

60. Lockwood, A. and Lockwood, A., 2022. *Animal Agriculture Was Critically Ignored at COP26—But Why?*. [online] Sentient Media. Available at: <https://sentientmedia.org/animal-agriculture-was-critically-ignored-at-cop26-but-why/> [Accessed 2 January 2022].

61. Ritchie, H. and Roser, M., 2022. *CO_2 and Greenhouse Gas Emissions*. [online] Our World in Data. Available at: <https://ourworldindata.org/greenhouse-gas-emissions> [Accessed 2 January 2022].

62. Ons.gov.uk. 2022. *Net zero and the different official measures of the UK's greenhouse gas emissions - Office for National Statistics*. [online] Available at: <https://www.ons.gov.uk/economy/environmentalaccounts/articles/netzeroandthedifferentofficialmeasuresoftheuksgreenhousegasemissions/2019-07-24> [Accessed 2 January 2022].

63. Our World in Data. 2022. *How do CO2 emissions compare when we adjust for trade?*. [online] Available at: <https://ourworldindata.org/consumption-based-co2> [Accessed 2 January 2022].

64. Our World in Data. 2022. *Who has contributed most to global CO2 emissions?*. [online] Available at: <https://ourworldindata.org/contributed-most-global-co2> [Accessed 2 January 2022].

65. the Guardian. 2022. *World's richest 1% cause double CO2 emissions of poorest 50%, says Oxfam*. [online] Available at: <https://www.theguardian.com/environment/2020/sep/21/worlds-richest-1-cause-double-co2-emissions-of-poorest-50-says-oxfam> [Accessed 2 January 2022].

66. Howrichami.givingwhatwecan.org. 2022. *How Rich Am I? World Income Percentile Calculator: Global Rich List*. [online] Available at: <https://howrichami.givingwhatwecan.org/how-rich-am-i> [Accessed 2 January 2022].

67. The Conversation. 2022. *Private planes, mansions and superyachts: What gives billionaires like Musk and Abramovich such a massive carbon footprint*. [online] Available at: <https://theconversation.com/private-planes-mansions-and-superyachts-what-gives-billionaires-like-musk-and-abramovich-such-a-massive-carbon-footprint-152514> [Accessed 2 January 2022].

68. 2022. [online] Available at: <https://twitter.com/neilrkaye/status/1452910090789572608> [Accessed 2 January 2022].

69. Unfccc.int. 2022. [online] Available at: <https://unfccc.int/news/climate-change-is-an-increasing-threat-to-africa> [Accessed 2 January 2022].

70. Unfccc.int. 2022. [online] Available at: <https://unfccc.int/files/press/backgrounders/application/pdf/factsheet_africa.pdf> [Accessed 2 January 2022].

71. Afrik 21. 2022. *AFRICA: The continent will have 105 million climate migrants by 2050 | Afrik 21*. [online] Available at: <https://www.afrik21.africa/en/africa-the-continent-will-have-105-million-climate-migrants-by-2050/> [Accessed 2 January 2022].

72. Torelli, S., 2022. *Climate-driven migration in Africa*. [online] ECFR. Available at: <https://ecfr.eu/article/commentary_climate_driven_migration_in_africa/> [Accessed 2 January 2022].

73. Ipcc.ch. 2022. [online] Available at: <https://www.ipcc.ch/site/assets/uploads/2018/02/AR5_SYR_FINAL_SPM.pdf> [Accessed 2 January 2022].

74. Greenpeace.org.uk. 2022. [online] Available at: <https://www.greenpeace.org.uk/news/climate-change-inequality-climate-justice/> [Accessed 2 January 2022].

75. CSIS Journalism Bootcamp. 2022. *Deforestation Hits Home: Indigenous Communities Fight for the Future of Their Amazon | CSIS Journalism Bootcamp*. [online] Available at: <https://journalism.csis.org/deforestation-hits-home-indigenous-communities-fight-for-the-future-of-their-amazon/> [Accessed 2 January 2022].

76. Worldometers.info. 2022. *World Population Clock: 7.9 Billion People (2022) - Worldometer*. [online] Available at: <https://www.worldometers.info/world-population/> [Accessed 2 January 2022].

77. Bettermeetsreality.com. 2022. *Ecological Footprint Of Each Country In The World - Better Meets Reality*. [online] Available at: <https://bettermeetsreality.com/ecological-footprint-of-each-country-how-many-earths-each-country-uses-based-on-consumption-rate-of-natural-resources/> [Accessed 2 January 2022].

78. livescience.com. 2022. *Greedy Nations: Top (and Bottom) Resource Users on Earth*. [online] Available at: <https://www.livescience.com/20308-greedy-nations-top-resource-users-earth.html> [Accessed 2 January 2022].

79. Unicef.org. 2022. *Girls' education*. [online] Available at: <https://www.unicef.org/education/girls-education> [Accessed 2 January 2022].

80. Climate & Migration Coalition. 2022. *Climate refugees: what is the definition? How is it decided?*. [online] Available at: <https://climatemigration.org.uk/climate-refugees-definition/> [Accessed 2 January 2022].

81. Hope, R., 2022. *Meet The Brexit Party's Climate Science Deniers*. [online] DeSmog. Available at: <https://www.desmog.com/2019/05/01/brexit-party-climate-science-deniers/> [Accessed 2 January 2022].

82. Data.worldbank.org. 2022. *Refugee population by country or territory of asylum | Data*. [online] Available at: <https://data.worldbank.org/indicator/SM.POP.REFG> [Accessed 3 January 2022].

83. International Committee of the Red Cross. 2022. *Seven things you need to know about climate change and conflict*. [online] Available at: <https://www.icrc.org/en/document/climate-change-and-conflict> [Accessed 3 January 2022].

84. 2022. [online] Available at: <https://www.sciencedaily.com/releases/2019/06/190620100005.htm > [Accessed 3 January 2022].

85. BERNERS-LEE, M., 2021. *HOW BAD ARE BANANAS?*. [S.l.]: GREYSTONE BOOKS.

86. Ritchie, H. and Roser, M., 2022. *CO_2 and Greenhouse Gas Emissions*. [online] Our World in Data. Available at: <https://ourworldindata.org/emissions-by-sector> [Accessed 22 January 2022].

87. CEOBS. 2022. *How does war damage the environment? - CEOBS*. [online] Available at: <https://ceobs.org/how-does-war-damage-the-environment/> [Accessed 3 January 2022].

88. WardheerNews. 2022. *This Is The Company Profiting Most From War - WardheerNews*. [online] Available at: <https://wardheernews.com/this-is-the-company-profiting-most-from-war/> [Accessed 9 January 2022].

89. Dorn, M., 2022. *Lockheed Martin – truthful politics*. [online] Truthfulpolitics.com. Available at: <http://www.truthfulpolitics.com/http:/truthfulpolitics.com/comments/tag/lockheed-martin/> [Accessed 9 January 2022].

90. Okaka, F. and Odhiambo, B., 2022. *Relationship between Flooding and Out Break of Infectious Diseasesin Kenya: A Review of the Literature*.

91. Nations, U., 2022. *Climate Change and Malaria - A Complex Relationship | United Nations*. [online] United Nations. Available at: <https://www.un.org/en/chronicle/article/climate-change-and-malaria-complex-relationship> [Accessed 3 January 2022].

92. Conservation.org. 2022. *Drivers and Causes of Zoonotic Diseases: An Overview*. [online] Available at: <https://www.conservation.org/research/drivers-and-causes-of-zoonotic-diseases-an-overview> [Accessed 3 January 2022].

93. Who.int. 2022. *WHO-convened global study of origins of SARS-CoV-2: China Part*. [online] Available at: <https://www.who.int/publications-detail-redirect/who-convened-global-study-of-origins-of-sars-cov-2-china-part> [Accessed 3 January 2022].

94. Clinical Microbiology Reviews. 2022. *Severe Acute Respiratory Syndrome Coronavirus as an Agent of Emerging and Reemerging Infection | Clinical Microbiology Reviews*. [online] Available at: <https://journals.asm.org/doi/10.1128/cmr.00023-07?permanently=true> [Accessed 3 January 2022].

95. Plastic Health Coalition. 2022. *How Plastic Affects & Threatens Human Health - Plastic Health Coalition*. [online] Available at: <https://www.plastichealthcoalition.org/> [Accessed 3 January 2022].

96. Sky News. 2022. *Human penises are shrinking because of pollution, warns scientist*. [online] Available at: <https://news.sky.com/story/human-penises-are-shrinking-because-of-pollution-warns-scientist-12255106> [Accessed 3 January 2022].

97. BBC News. 2022. *Climate change: Young people very worried - survey*. [online] Available at: <https://www.bbc.co.uk/news/world-58549373> [Accessed 3 January 2022].

98. Dissertations expert. 2022. *Education is the most powerful weapon nelson mandela*. [online] Available at: <https://dissertationsexperts.com/education/education-is-the-most-powerful-weapon-nelson-mandela.html> [Accessed 3 January 2022].

99. the Guardian. 2022. *Owen Paterson v the science of climate change*. [online] Available at: <https://www.theguardian.com/environment/blog/2013/sep/30/owen-paterson-science-climate-change> [Accessed 4 January 2022].

100. Yale Program on Climate Change Communication. 2022. *School's starting: A majority of Americans say global warming should be taught in the classroom. - Yale Program on Climate Change Communication*. [online] Available at: <https://climatecommunication.yale.edu/news-events/schools-starting-majority-americans-say-global-warming-taught-classroom/> [Accessed 4 January 2022].

101. Teachthefuture.uk. 2022. *Teach the Future: Campaign for climate education*. [online] Available at: <https://www.teachthefuture.uk/> [Accessed 4 January 2022].

102. HISTORY. 2022. *How Nelson Mandela Used Rugby as a Symbol of South African Unity*. [online] Available at: <https://www.history.com/news/nelson-mandela-1995-rugby-world-cup-south-african-unity> [Accessed 5 January 2022].

103. BBC Sport. 2022. *Man Utd defend 100-mile Leicester flight*. [online] Available at: <https://www.bbc.co.uk/sport/football/58938198> [Accessed 5 January 2022].

104. the Guardian. 2022. *Sampdoria's Morten Thorsby: 'They call me football's Greta Thunberg'*. [online] Available at: <https://www.theguardian.com/football/2021/dec/16/sampdoria-

morten-thorsby-they-call-me-football-greta-thunberg> [Accessed 5 January 2022].

105. The Odyssey Online. 2022. *12 Protests That Changed History*. [online] Available at: <https://www.theodysseyonline.com/protests-that-changed-history> [Accessed 5 January 2022].

106. Vesty, H., 2022. *Ambulances wait outside for '12 hours' as hospitals 'not allowing patients in'*. [online] Manchester Evening News. Available at: <https://www.manchestereveningnews.co.uk/news/greater-manchester-news/ambulances-waiting-outside-12-hours-21968251> [Accessed 9 January 2022].

107. openDemocracy. 2022. *British MPs urge privacy watchdog to investigate BP for 'political spying'*. [online] Available at: <https://www.opendemocracy.net/en/opendemocracyuk/british-mps-urge-privacy-watchdog-to-investigate-bp-for-political-spying/> [Accessed 22 January 2022].

108. Stacker. 2022. *Biggest corporate subsidies of the last 20 years*. [online] Available at: <https://stacker.com/stories/2946/biggest-corporate-subsidies-last-20-years> [Accessed 12 January 2022].

109. Citizens' Climate Lobby UK. 2022. *Climate Income | Citizens' Climate Lobby UK*. [online] Available at: <https://citizensclimatelobby.uk/climate-income/> [Accessed 12 January 2022].

110. Psychology Today. 2022. *Influencing Wellbeing: Materialism and Self-Esteem*. [online] Available at: <https://www.psychologytoday.com/gb/blog/in-the-neighborhood/201403/influencing-wellbeing-materialism-and-self-esteem> [Accessed 13 January 2022].

Chapter 9: So, What can I do?

1. David G. Allan, C., 2021. *Breaking news: People are inherently good, nonviolent.* [online] CNN. Available at: <https://edition.cnn.com/2018/12/11/health/nonviolence-good-wisdom-project/index.html> [Accessed 7 November 2021].

2. Jeon, K., 2021. *The Life and Culture of the Bajau, Sea Gypsies*.

Printed in Great Britain
by Amazon

81283922R00244